ADVANCES IN SATELLITE COMMUNICATIONS

Edited by **Masoumeh Karimi** and **Yuri Labrador**

INTECHOPEN.COM

Advances in Satellite Communications

Edited by Masoumeh Karimi and Yuri Labrador

CBS, Edition **2016**

Published by InTech

Janeza Trdine 9, 51000 Rijeka, Croatia

Publishing Process Manager Mia Devic

Technical Editor Teodora Smiljanic

Cover Designer InTech Design Team

Image Copyright Chaikovskiy Igor, 2010. Used under license from Shutterstock.com

Additional hard copies can be obtained from orders@intechopen.com

Advances in Satellite Communications, Edited by Masoumeh Karimi and Yuri Labrador
p. cm.
ISBN 978-953-307-562-4

Contents

Preface

The use of satellites in communications systems is very much a fact of everyday life, as is evidenced by the many homes equipped with antennas, or dishes used for reception of satellite television. What may not be so well known is that satellites form an essential part of telecommunication systems worldwide, carrying large amounts of data and telephone traffic in addition to the television signals.

Satellite communication systems are now a major part of most telecommunications networks as well as our everyday lives through mobile personal communication systems and broadcast television. A sound understanding of such systems is therefore important for a wide range of system designers, engineers and users.

Satellites offer a number of features not readily available with other means of communications. Because very large areas of the earth are visible from a satellite, the satellite can form the star point of communications network, simultaneously linking many who may be widely separated geographically. The same features enable satellites to provide communication links to remote communities in sparsely populated areas that are difficult to access by other means. Of course, satellites' signals ignore political boundaries as well as geographic ones, which may or may not be a desirable feature.

A satellite communications system uses a variety of technologies combining many areas of engineering fields, from video compression to access techniques, modulation schemes, and error correction capabilities. In addition, a satellite link engineer has to consider the constrains of the communication channel; so many radio frequency calculations are needed, these include link budget analysis, propagation effects, amplifiers' operational points, antennas size, transmission lines, impedance matching, etc.

This book provides a comprehensive review of some applications that have driven this growth. It analyzes various aspects of Satellite Communications from Antenna design, Real Time applications, Quality of Service (QoS), Atmospheric effects, Hybrid Satellite-Terrestrial Networks, Sensor Networks and High Capacity Satellite Links.

It is the desire of the editors that the topics selected for this book can give the reader an overview of the current trends in Satellite Systems, and also an in depth analysis of the technical aspects of each one of them. The editors would like to acknowledge the support of all of those people at the InTech (Open Access Publisher) who have helped during the process of this book. In particular, special thanks to Ms. Mia Devic for her valuable comments, hard work and continued guidance.

Dr Masoumeh Karimi
Technological University of America,
USA

Dr Yuri Labrador
The MITRE Corporation,
USA

Part 1

Antennas in Satellite Communications

1

Helical Antennas in Satellite Radio Channel

Maja Škiljo and Zoran Blažević
University of Split, Faculty of electrical engineering,
mechanical engineering and naval architecture,
Croatia

1. Introduction

Monofilar and multifilar helical antennas are the most widely proposed antennas in satellite communications systems. The main reason why these antennas constitute an asset in applications concerning satellite and space communications generally is circular polarization. Good axial ratio provides precise measurement of the polarization of the received signal due to immunity of the circularly polarized wave to Faraday rotation of the signal propagating through the ionosphere.

In addition to circular polarization, monofilar helical antennas offer the advantage of high gain in axial direction over a wide range of frequencies which makes them suitable for applications in broadband satellite communications. Split beam and conical beam radiation patterns of bifilar and quadrifilar helical antennas respectively, offer even more applications in mobile satellite communications (Kilgus, 1975; Nakano et al., 1991). Also, backfire helical antenna has stood out as a better feed element for parabolic reflector than the axial mode helical antenna and horn antennas (Nakano et al., 1988). Beside the number of wires in helical antenna structure, it is possible to use antenna's physical parameters to control the directivity pattern. Phase velocity of the current can be controlled by changing the pitch angle and circumference (Kraus, 1988; Mimaki & Nakano, 1998), and the ground plane can be varied in its size and shape to achieve a certain form of radiation pattern and higher antenna gain (Djordjevic et al., 2006; Nakano et al., 1988; Olcan et al., 2006). Various materials used in helical antenna design, even only for the purpose of mechanical support or isolation, can noticeably influence the antenna's performance so this should be taken into account when designing and modeling the desirable helical antenna structure (Casey & Basal, 1988a; Casey & Basal, 1988b; Hui et al., 1997; Neureuther et al., 1967; Shestopalov et al., 1961; Vaughan & Andersen, 1985).

A theoretical study of a sheath, tape and wire helix given in (Sensiper, 1951) provided the base for a physical model of the helical antenna radiation mechanism. The complex solutions of the determinantal equation for the propagation constants of the surface waves traversing a finite tape helix are used to calculate the current distribution on helical antenna in (Klock, 1963). The understanding of the waves propagating on the helical antenna structure can also provide a good assessment of the circular polarization purity as well as the estimation of varying the helical antenna radiation characteristics by changing the antenna's physical parameters and using various materials in helical antenna design (Maclean & Kouyoumjian, 1959; Neureuther et al., 1967; Vaughan & Andersen, 1985).

Although an analytical approach can sometimes provide a fast approximation of helix radiation properties (Maclean & Kouyoumjian, 1959), generally it is a very complicated procedure for an engineer to apply efficiently and promptly to the specified helical antenna design. Therefore, we combine the analytical with the numerical approach, i. e. the thorough understanding of the wave propagation on helix structure with an efficient calculation tool, in order to obtain the best method for analyzing the helical antenna.

In this chapter, a theoretical analysis of monofilar helical antenna is given based on the tape helix model and the antenna array theory. Some methods of changing and improving the monofilar helical radiation characteristics are presented as well as the impact of dielectric materials on helical antenna radiation pattern. Additionally, backfire radiation mode formed by different sizes of a ground reflector is presented. The next part is dealing with theoretical description of bifilar and quadrifilar helices which is followed by some practical examples of these antennas and matching solutions. The chapter is concluded with the comparison of these antennas and their application in satellite communications.

2. Monofilar helical antennas

The helical antenna was invented by Kraus in 1946 whose work provided semi-empirical design formulas for input impedance, bandwidth, main beam shape, gain and axial ratio based on a large number of measurements and the antenna array theory. In addition, the approximate graphical solution in (Maclean & Kouyoumjian, 1959) offers a rough but also a fast estimation of helical antenna bandwidth in axial radiation mode. The conclusions in (Djordjevic et al., 2006) established optimum parameters for helical antenna design and revealed the influence of the wire radius on antenna radiation properties. The optimization of a helical antenna design was accomplished by a great number of computations of various antenna parameters providing straightforward rules for a simple helical antenna design.

Except for the conventional design, the monofilar helical antenna offers many various modifications governed by geometry (Adekola et al., 2009; Kraft & Monich, 1990; Nakano et al., 1986; Wong & King, 1979), the size and shape of reflector (Carver, 1967; Djordjevic et al., 2006; Nakano et al., 1988; Olcan et al., 2006), the shape of windings (Barts & Stutzman, 1997, Safavi-Naeini & Ramahi, 2008), the various guiding (and supporting) structures added (Casey & Basal, 1988a; Casey & Basal, 1988b; Hui et al., 1997; Neureuther et al., 1967; Shestopalov et al., 1961; Vaughan & Andersen, 1985) and other. This variety of multiple possibilities to slightly modify the basic design and still obtain a helical antenna performance of great radiation properties with numerous applications is the motivation behind the great number of helical antenna studies worldwide.

2.1 Helix as an antenna array

A simple helical antenna configuration, consisted of a perfectly conducting helical conductor wounded around the imaginary cylinder of a radius a with some pitch angle ψ, is shown in Fig. 1. The conductor is assumed to be a flat tape of an infinitesimal thickness in the radial direction and a narrow width δ in the azimuthally direction. The antenna geometry is described with the following parameters: circumference of helix $C = \pi D$, spacing p between the successive turns, diameter of helix $D = 2a$, pitch angle $\psi = \tan^{-1}(p/\pi D)$, number of turns N, total length of the antenna $L = Np$, total length of the wire $L_n = NL_0$ where L_0 is the wire length of one turn $L_0 = (C^2 + p^2)^{1/2}$.

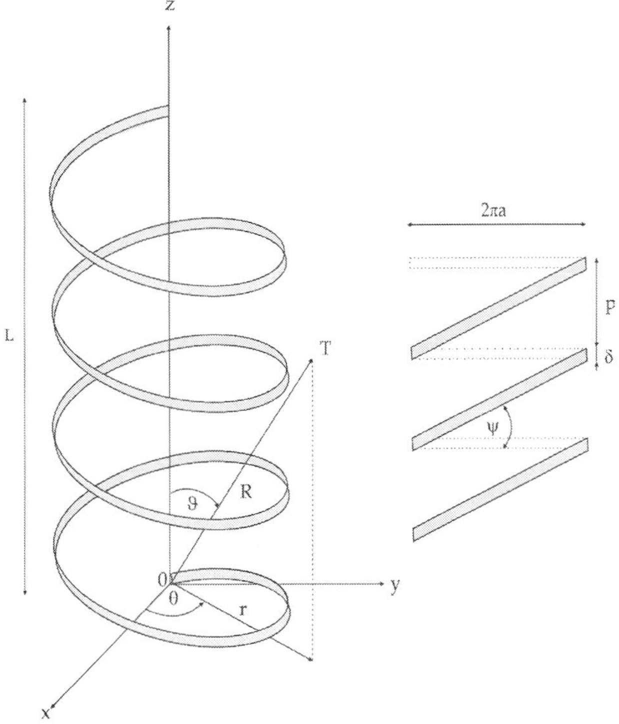

Fig. 1. The tape helix configuration and the developed helix.

Considering the tape is narrow, $\delta << \lambda$, p, a, assuming the existence of electric and magnetic currents in the direction of the antenna axis of symmetry and applying the boundary conditions on the surface of the helix, we can derive the field expressions for each existing free mode as the total of an infinite number of space harmonics caused by helix periodicity with the propagation constants $h_m = h + 2\pi m/p$, where m is an integer (Sensiper, 1951). Knowing the field components at the antenna surface, the far field in spherical coordinates (R, θ, ϑ) for each existing mode can be obtained upon by the Kirchhoff-Huygens method. The contribution to the radiated field of each space harmonic can be written in the form of the element factor and the array factor product, thus the total radiated electric field caused by the particular mode is expressed as (Cha, 1972; Kraus, 1948; Shestopalov, 1961; Vaughan & Andersen, 1985):

$$E_\theta(\theta,\vartheta) = \sum_{m=-\infty}^{\infty} F_{\theta m}(\theta,\vartheta) G_m(\vartheta;L), \tag{1}$$

$$E_\vartheta(\theta,\vartheta) = \sum_{m=-\infty}^{+\infty} F_{\vartheta m}(\theta,\vartheta) G_m(\vartheta;L). \tag{2}$$

The element factors $F_{\theta m}$ and $F_{\vartheta m}$ represent the contribution of each turn to the total field in some far point of the space due to the mth cylindrical space harmonic, and are determined as:

$$F_{\theta m}(\theta,\vartheta) = 2\left(\frac{m}{ka}E^a_{zm}\cot\vartheta - E^a_{\theta m}\sin\vartheta\right)J_m - jZ_0 H^a_{zm}\left(J_{m+1} - J_{m-1}\right), \tag{3}$$

$$F_{\vartheta m}(\theta,\vartheta) = 2Z_0\left(\frac{m}{ka}H^a_{zm}\cot\vartheta - H^a_{\theta m}\sin\vartheta\right)J_m + jE^a_{zm}\left(J_{m+1} - J_{m-1}\right), \tag{4}$$

where $E^a_{\theta m}$, $E^a_{\vartheta m}$, and $H^a_{\theta m}$, $H^a_{\vartheta m}$ are the mth cylindrical space harmonic amplitudes of electric and magnetic field spherical components at the antenna surface respectively, $k = 2\pi f\sqrt{\mu_0\varepsilon_0} = 2\pi f/c$ is the free-space wave-number, $Z_0 = \sqrt{\mu_0/\varepsilon_0} = 120\pi\ \Omega$ is the impedance of the free space, and $J_m = J_m(ka\sin\vartheta)$ is the ordinary Bessel function of the first kind and order m. The complex array factor G_m is calculated for each space harmonic as:

$$G_m(\vartheta;L) = L\,\text{sinc}\left(N\,\Phi_m/2\right)e^{jN\Phi_m/2}, \tag{5}$$

where Φ_m is the phase difference for the mth harmonic between the successive turns:

$$\Phi_m = kL\left(\cos\vartheta - \frac{h_m}{k}\right). \tag{6}$$

Unlike the element factor, the array factor defines the directivity and does not influence the polarization properties of the antenna. It is found (Kraus, 1949) that, although (3) and (4) are different in form, the patterns (1) and (2) for entire helix are nearly the same, and the similar could also be stated for the dielectrically loaded antenna. Furthermore, the main lobes of E_θ and E_ϑ patterns are very similar to the array factor pattern. Hence, the calculation of the array factor alone suffices for estimations of the antenna properties at least for long helices. Assuming only a single travelling wave on the helical conductor, following (1)-(2), a helix antenna can be depicted as an array of isotropic point sources separated by the distance p, as in Fig. 2. The normalized array factor is:

$$G_A = \frac{\sin\left(N\,\Phi/2\right)}{N\sin\left(\Phi/2\right)}. \tag{7}$$

This is justified as the absolute of (5) and (7) are approximately equal, and small differences become noticeable only for $N \le 5$. Denoting the phase difference for the fundamental space harmonic of axial mode as $\Phi_0 = \Phi$ in (6), the Hansen-Woodyard condition for the maximum directivity in the axial direction ($\vartheta = 0$) states that (Maclean & Kouyoumjian, 1959):

$$\Phi = -2\pi\left(1 + \frac{1}{2N}\right), \tag{8}$$

Ideally, applying (6)-(8), the radiation characteristics of the helical antenna and the antenna geometry can be directly connected by single variable, the velocity v of the surface wave (Kraus, 1949; Maclean & Kouyoumjian, 1959; Nakano et al., 1986; Wong & King, 1979). As the wave velocities in a finite helix are hard to calculate, those calculated for the infinite

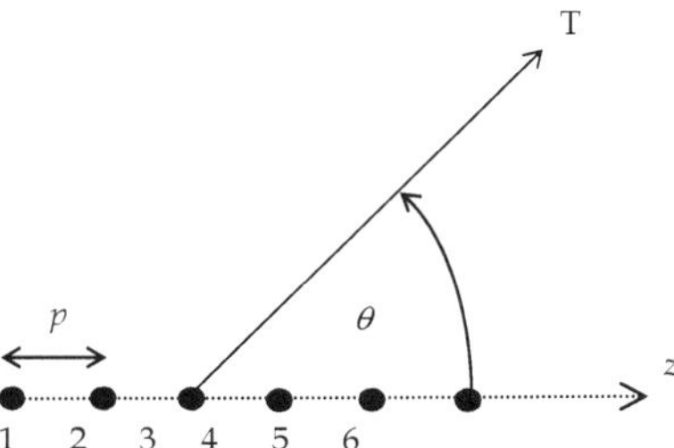

Fig. 2. The array of N point sources.

helix can be applied as a fair approximation. The determinantal equation for the wave propagation constants on an infinite helical waveguide is given and analyzed in (Klock, 1963; Mittra, 1963; Sensiper, 1951, 1955) and generalized forms of the equation for helices filled with dielectrics are considered in (Blazevic & Skiljo, 2010; Shestopalov et al. 1961; Vaughan & Andersen, 1985). The solutions are obtained in a form of the Brillouin diagram for periodic structures, which dispersion curves are symmetrical with respect to the ordinate (the circumference of the helix in wavelengths). The calculated propagation constants (phase velocities) of free modes are real numbers settled within the triangles defined by lines $ka = \pm ha \mp |m| \cot\psi$, among which those with $|m| = 1$ comply with the condition (8) for infinite arrays. The $m = 0$ and $m = -1$ regions of the diagram refer to the so called normal and the axial mode, respectively. The Brillouin diagram provide the information about the group velocity of the surface waves calculated as the slope of the dispersion curves at given frequency. It is important to note that the phase and group velocities on the helix may have opposite directions. When the circumference of the helix is small compared to the wavelength, the normal mode dominates over the others and the maximum radiated field is perpendicular to helix axis. These electric field components are then out of phase so the total far field is usually elliptically polarized. Due to the narrow bandwidth of radiation, the normal mode helical antenna is limited to narrow band applications (Kraus, 1988). Axial radiation mode is obtained when the circumference of helix is approximately one wavelength, achieving a constructive interference of waves from the opposite sides of turns and creating the maximum radiation along the axis. Helical antenna in the axial mode of radiation is a circularly polarized travelling-wave wideband antenna.

However, due to the assumption of the existence of only a single travelling wave, the modeling of helical antenna as a finite length section of the helical waveguide has some practical shortcomings, which becomes more problematical as the antenna length becomes shorter. Consider an example of the typical axial mode current distribution on Fig. 3, obtained at $C_\lambda = 1.0$ for the helical antenna with $\psi = 14°$ and $N = 12$. We may observe three regions: the exponential decaying region away from the source, the surface wave region after the first minimum and the standing wave due to reflection of the outgoing wave at the open antenna end. The works of (Klock, 1963; Kraus, 1948, 1949; Marsh, 1950) showed that the approximate current distribution can be estimated assuming two main current waves, one with a complex valued phase constant settled in the region of normal mode ($m = 0$) that forms a standing wave deteriorating antenna radiation pattern, and one with real phase constant in the region of the axial mode ($m = -1$) that contributes to the beam radiation.

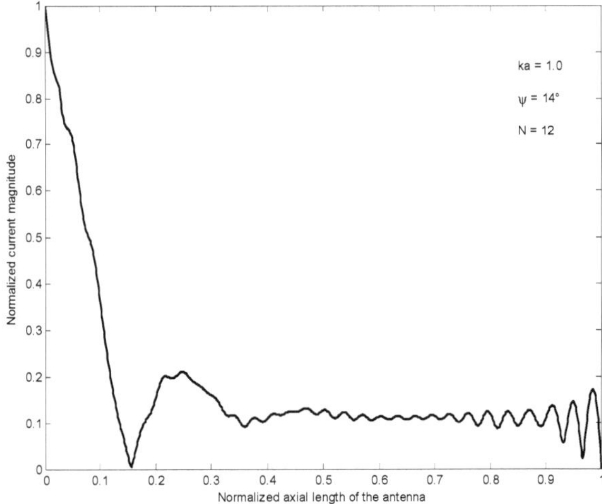

Fig. 3. A typical axial mode current distribution on helical antenna.

The analytical procedure of a satisfying accuracy for determining the relationship between the powers of the surface waves traversing the arbitrary sized helical antenna may still be sought using a variational technique, assuming the existence of only two principal propagation modes (normal and axial), and a sinusoidal current distributions for each of them taking into account the velocities calculated for the infinite helical waveguide, as shown by (Klock, 1963). However, as the formula for the total current on the helix involves integrals of a very complex form, one may rather chose to use the classical design data given in (Kraus, 1988) which, for helices longer than three turns, define the optimum design parameters in a limited span of the pitch angles in the frequency range of the axial mode. The semi-empirical formulas for antenna gain G in dB, input impedance R in ohms, half power beam-width $HPBW$ in degrees and axial ratio AR, are given by:

$$G = 11.8 + 10\log\left(N\left(\frac{p}{\lambda}\right)\left(\frac{C}{\lambda}\right)^2\right), \tag{9}$$

$$R = 140\frac{C}{\lambda}, \tag{10}$$

$$AR = \frac{2N+1}{2N}, \tag{11}$$

$$HPBW = \frac{52}{\left(\frac{C}{\lambda}\right)\sqrt{N\left(\frac{p}{\lambda}\right)}}. \tag{12}$$

Because of the traveling-wave nature of the axial-mode helical antenna, the input impedance is mainly resistive and frequency insensitive over a wide bandwidth of the antenna and can be estimated by (10). The discrepancy from a pure circular polarization, described with axial ratio AR, depends on the number of turns N and it approaches to unity as the number of turns increases. It is interesting to note that this formula is obtained by Kraus using a quasi-empirical approach where the phase velocity is assumed to always satisfy the Hansen-Woodyard condition for increased directivity. The reflected current degrades desired polarization in forward direction and by suppressing it (with tapered end for example); the formula (11) becomes more accurate (Vaughan & Andersen, 1985). However, King and Wong reported that without the end tapering the axial ratio formula often fails (Wong & King, 1982). Also, based on a great number of experimental results, they established that in the equation (13), valid for $12° < \psi < 15°$, $3/4 < C/\lambda < 4/3$ and $N > 3$, numerical factor can be much lower than 15, usually between 4.2 and 7.7 (Djordjevic et al., 2006), providing a different expression for the helical antenna gain:

$$G = 8.3 \left(\frac{\pi D}{\lambda_p} \right)^{\sqrt{N+2}-1} \left(\frac{Np}{\lambda_p} \right)^{0.8} \left[\frac{\tan 12.5°}{\tan \psi} \right]^{\frac{\sqrt{N}}{2}}, \tag{13}$$

where λ_p is wavelength at peak gain.

The existence of multiple free modes on a helical antenna makes the theoretical analysis even more complicated when a dielectric loading is introduced. Consider two examples of the Brillouin diagram in the region $m = -1$ for the case of $\psi = 13°$, $\delta = 1$ mm, $N = 10$ given on Fig. 4 a) and b) respectively. The first refers to the empty helix and the second to the helix filled uniformly with a lossless dielectric of relative permittivity $\varepsilon_r = 6$. The A points mark the intersections of the dispersion curves of the determinantal equation with the line defined by the Hansen-Woodyard condition (8). Obviously, their positions depend on the number of turns. Point B marks the calculated upper frequency limit of the axial mode, f_B i.e. the frequency at which the SLL is increased to 45 % of the main beam, the criterion adopted from (Maclean & Kouyoumjian, 1959). In the case of helical antenna with dielectric core, due to the difference in permittivity of the antenna core and surrounding media, it can be noted that the solutions shape multiple branches. It can also be shown that the number of branches increases rapidly by increasing the permittivity and decreasing the pitch angle.

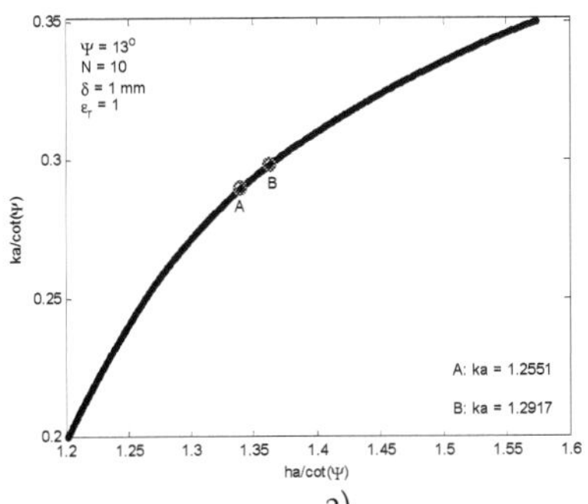

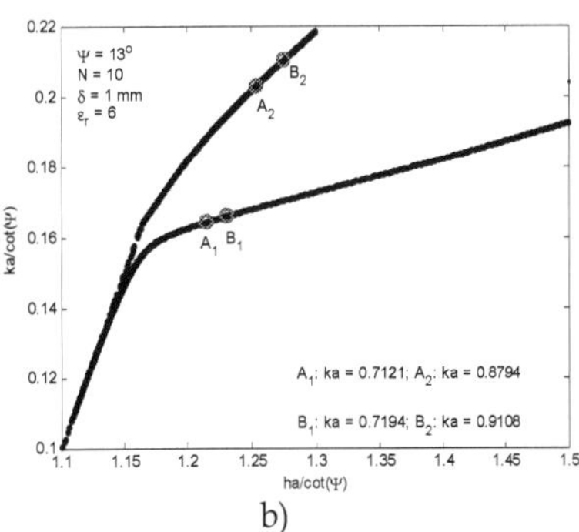

Fig. 4. A section of the Brillouin diagram in the axial mode region ($m = -1$) for the tape helix with parameters $\psi = 13°$, $\delta = 1$ mm, $N = 10$, $\varepsilon_r = 1$ a) and $\varepsilon_r = 6$ b).

The existence of multiple axial modes as in Fig. 4 b) implicates a possibility of the existence of a number of optimal frequencies (A points), one for each axial mode. However, if the permittivity is high enough and the pitch angle low enough, the power of the lowest axial mode may be found to be insufficient to shape a significant beam radiation. Then the solution A at the lowest mode branch of the dispersion curve is settled below the minimum beam mode frequency f_L. This frequency limit marks the frequency at which the axial mode power starts to dominate over the normal mode power. It is usually determined as the lowest frequency at which the circular polarization is formed i.e. the axial ratio is less than two. Also, the HPBW of the main lobe falls below 60 degrees but this criterion can be strictly applied only for longer helices (longer than ten turns). As the working frequency starts to surmount this limit, the current magnitude distribution is transformed steadily toward the classical shape of the axial mode current (Kraus, 1988) as in Fig. 3. Also, as the classical current distribution forms, the character of the input impedance starts to be mainly real. It is found in (Maclean & Kouyoumjian, 1959) that the lower limit remains approximately constant regardless of the antenna length. This fact is confirmed for the dielectrically loaded helices as well in (Blazevic & Skiljo, 2010). It is also noted that the change in the maximum axial mode frequency with varying permittivity and pitch angle as the consequence of the change of the surface wave group velocity is much more emphasized than the change of the minimum frequency. This means that, as the optimal frequency becomes lower, the axial mode bandwidth shrinks. The overall effect of the permittivity and pitch angle on the fractional axial mode bandwidth (defined as the ratio of the bandwidth and twice the central frequency) for the various antenna lengths is depicted on Fig. 5.

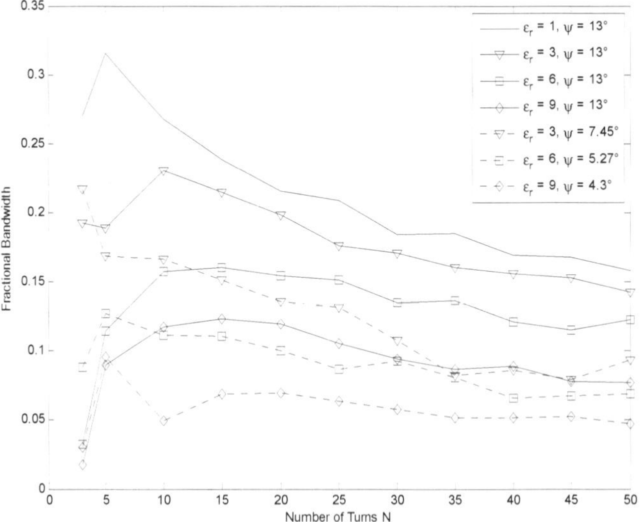

Fig. 5. The axial mode fractional bandwidth of the antennas for various dielectric loadings and pitch angles vs. number of turns.

2.2 Impact of materials used in helical antenna design

A frequently used antenna is the conventional monofilar helical structure wrapped around a hollow dielectric cylinder providing a good mechanical support, especially for thin and long helical antennas. In the case of commercially manufactured helical antennas they are often covered with non-loss dielectric material all over, while in amateur applications sometimes low cost lossy materials take place. The properties of various materials used in antenna design and their selection can be of great importance for meeting the required antenna performance, and the purpose of this chapter is to provide an insight to its influence based on a practical example.

The CST Microwave Studio was used to analyze the impact of various materials and their composition on helical antenna design and optimal performance. Since the chapter focuses on longer antennas, a 12-turn helix was chosen. We created the helical structure with the following parameters: f = 2430 MHz, D = 42 mm, C = 132 mm, p = 33 mm, L = 396 mm, N = 12, a = 1 mm and Ψ = 14°. Instead of infinite ground plane commonly used in numerical simulations, we formed a round reflector with the diameter of D_r = 17 cm to be closer to the widespread practical design. The resistance of the source is selected to be 50 Ω and the thickness of the dielectric tube in practical design is 1mm.

The antenna shown in Fig. 6 a) is the reference model of the helical antenna constructed of a perfectly conducting helical conductor and a finite size circular reflector using the hexahedral mesh.

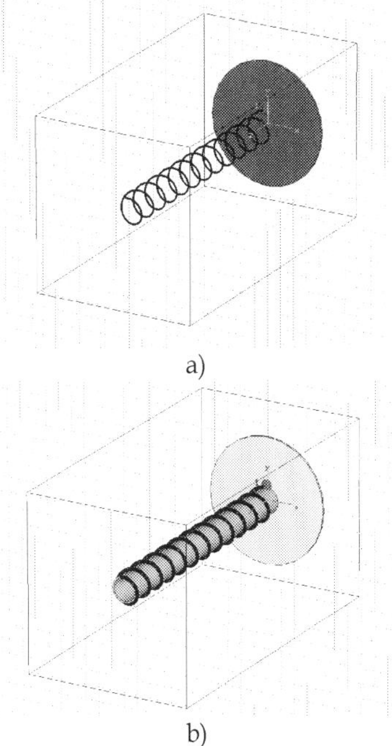

a)

b)

Fig. 6. The simulated helical antenna structures: a) the reference model and b) the practical design simulation.

The simulation results in Fig. 7 demonstrate the influence of applied materials on the antenna VSWR and gain in frequency band from 1.8-2.8 GHz. Each material was examined separately except for the practical design of the antenna which included all the materials used. First step to practical design of the helical antenna depicted in Fig. 6 a) was the replacement of the PEC material with the copper one, which produced negligible effects on the antenna parameters as expected. Lossy dielectric wire coating added to reference model with permittivity and conductivity selected to be $\varepsilon_r = 3$ and $\sigma = 0.03$ S/m, however, caused noticeable change in the overall antenna performance. The antenna input impedance is decreased where primarily the capacitive reactance is decreased because of the higher permittivity along the helical conductor. Also, the gain is decreased and the frequency bandwidth of the antenna is shifted to somewhat lower frequencies. The empty dielectric tube (EDT), often used as a mechanical support for long antennas, is analyzed in two steps. First, non-loss EDT (with $\varepsilon_r = 3$) added to the reference model, produced gain decrease and the bandwidth shift. At the same time, the antenna input impedance decreases causing the improvement of VSWR. When the conductivity of $\sigma = 0.03$ S/m is added in second step, these effects are much more emphasized, especially for the antenna gain.

Comparing the obtained antenna gain of 13.96 dB at $f = 2.43$ GHz of reference PEC model with (9) and (13), where calculated gains are $G = 17.44$ dB and $G = 13.21$ dB respectively, it is found that the first formula is too optimistic as expected, and the second one is acceptable for some readily estimation of helical antenna gain. To the reference, the final practical antenna design, comprising the copper helical wire covered with lossy dielectric wire coating wounded around the lossy dielectric tube, and the finite size circular reflector, achieves gain of 10.91 dB at 2.43 GHz and peak gain of 13.18 dB at 2.2 GHz. Thus, in comparison with PEC helical antenna in free space, the practical antenna performance is significantly influenced by the dielectric coating and supporting EDT.

2.3 Changing the parameters of helix to achieve better radiation characteristics

High antenna gain and good axial ratio over a broad frequency band are easily achieved by various designs of a helical antenna which can take many forms by varying the pitch angle (Mimaki and Nakano, 1998; Nakano et al., 1991; Sultan et al., 1984), the surrounding medium (Bulgakov et al., 1960; Casey and Basal, 1988; Vaughan and Andersen, 1985) and the size and shape of reflector (Djordjevic et al., 2006; Nakano et al., 1988; Olcan et al., 2006). In this chapter, we introduce a design of the helical antenna obtained by combining two methods to improve the radiation properties of this antenna; one is changing the pitch angle, i.e. combining two pitch angles (Mimaki and Nakano, 1998; Sultan et al., 1984) and the other is reshaping the round reflector into a truncated cone reflector (Djordjevic et al., 2006; Olcan et al., 2006).

It is shown (Mimaki and Nakano, 1998) that double pitch helical antenna radiates in endfire mode with slightly higher gain over wider bandwidth. Two pitch angles were investigated; 2° and 12.5°, along different lengths of the antenna. Their relative lengths were varied in order to obtain a wider bandwidth with higher antenna gain. In (Skiljo et al., 2010) the axial mode bandwidth was examined by means of parameters defining the limits of the axial radiation mode: axial ratio, HPBW, side lobe level (SLL) and total gain in axial direction, whereas the method of changing the pitch angle was applied to a helical antenna wounded around a hollow dielectric cylinder with the pitch angle of 14°. The maximum gain of the antennas with variable lengths h/H, where h is the antenna length where pitch angle $\psi_h = 2°$

and H is the rest of the antenna with $\psi_H = 12.5°$, is achieved with $h/H = 0.26$ (Mimaki and Nakano, 1998; Skiljo et al., 2010).

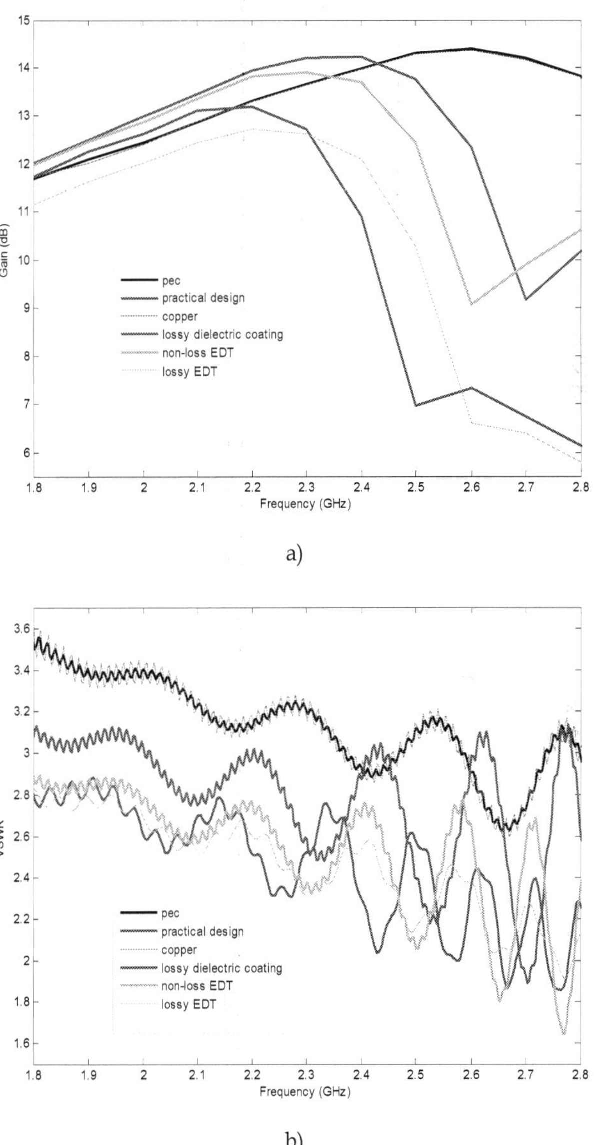

a)

b)

Fig. 7. The simulation results of material influence on antenna a) gain and b) VSWR.

Various shapes of ground plane were considered: infinite ground plane, square conductor, cylindrical cup and truncated cone, whereas the later produced the highest gain increase relative to the infinite ground plane. So, we used the truncated cone reflector with optimal cone diameters $D_1 = 1.3\lambda$ and $D_2 = 0.4\lambda$ and height $h = 0.5\lambda$ in order to maximize the gain of the previously simulated double pitch helical antenna (Skiljo et al., 2010). Applying the criteria for the cut-off frequencies of the axial mode from chapter 2.1, it is observed that the bandwidth of the axial mode is not increased (it is slightly shifted towards lower frequencies) by using two pitch angles and a truncated cone reflector. Fig 8 a) shows the antenna model used in chapter 2.2 with non loss dielectric tube (with $\varepsilon_r = 3$) and b) the simulated double pitch helical antenna.

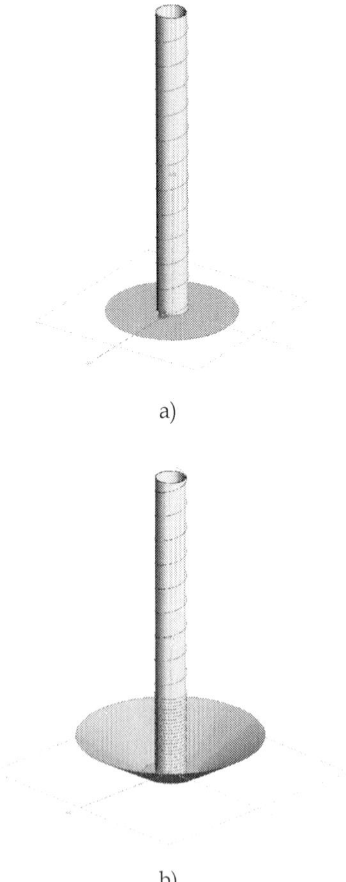

a)

b)

Fig. 8. Simulation of the a) standard twelve turn helical antenna and b) double pitch helical antenna with truncated cone reflector.

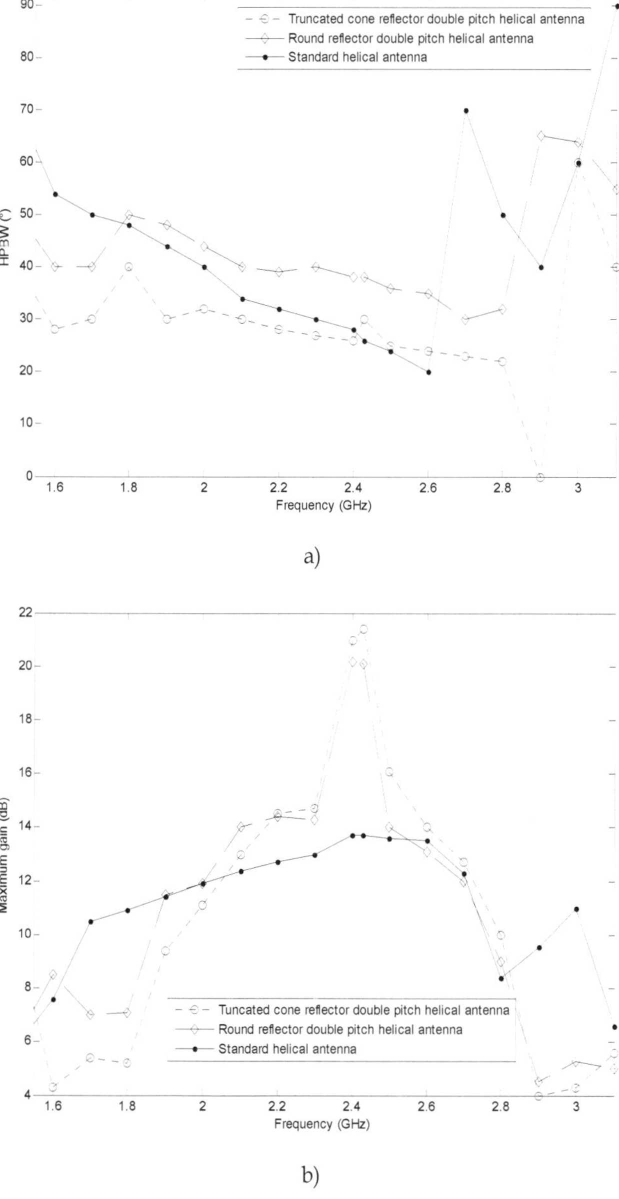

a)

b)

Fig. 9. a) HPBW and b) total antenna gain comparison between the standard twelve turn helical antenna, double pitch helical antenna with truncated cone, and with round reflector.

The results in Fig. 9 depict that HPBW is mainly better in case of the truncated cone reflector but worse with the round reflector, and the antenna gain is improved when using the truncated cone. Also, Fig. 9 b) shows a significant gain increase of the double pitch helical antenna with truncated cone reflector in comparison with the standard one around 2.4 GHz, but the bandwidth of such an antenna gain is not increased.

2.4 Backfire monofilar helical antenna

This chapter gives the information about the effect of the ground plane size on the helical antenna radiation characteristics. It is found that as the diameter of the reflector decreases, the backfire radiation occurs and at the ground plane diameter smaller than the helix diameter it becomes dominant (Nakano et al., 1988). The analysis of a monofilar backfire helix was carried out through the example from chapter 2.1: $\lambda = 12.34$ cm, $\psi = 14°$, $N = 12$, $r_w = 0.008\lambda$ and $D = 0.34\lambda$ with the reflector diameter of $d = 1.38\lambda$. This antenna can also be used in the form of monofilar backfire helix in the focus of a paraboloidal reflector. The results of simulations performed in FEKO show the radiation patterns and current distributions of the helical antennas with three different diameters of ground plane $d_1 = 0.7\lambda$, $d_2 = 0.35\lambda$ and $d_3 = 0.3\lambda$. In Fig. 10 a) helical antenna operates in standard axial mode where radiation is in forward direction where relative phase velocity $p = v/c$ satisfies the in-phase Hansen-Woodyard condition and the current distribution shows that surface wave is formed after the first minimum. There are no great discrepancies between this antenna and the one with larger reflector, as expected. As the diameter of the reflector decreases below 0.5λ, the decaying region of current distribution (Fig. 10) slightly shifts toward the end and becomes comparable to the surface region of the current. Also the amplitude of current in surface wave region decreases meaning that the backward radiation becomes larger. The antenna in Fig. 10 c) is the typical backfire monofilar helical antenna with the current distribution consisted only of a decaying current and a relative phase velocity nearly equal to one. It can be noticed that the forward and backward wave helical antennas achieve good but opposite sense circular polarization (Nakano et al., 1988).

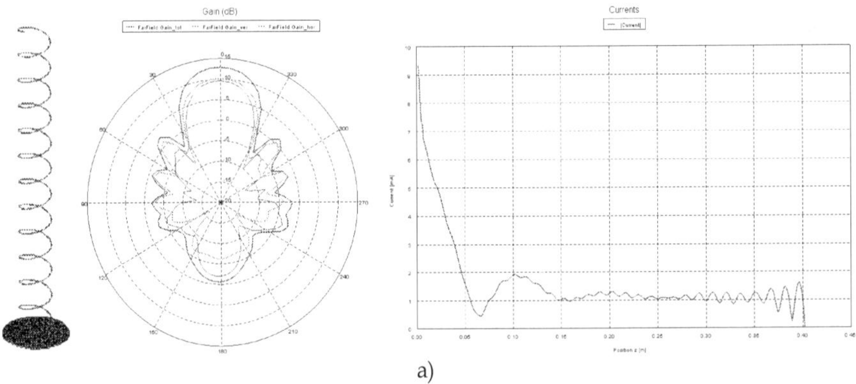

a)

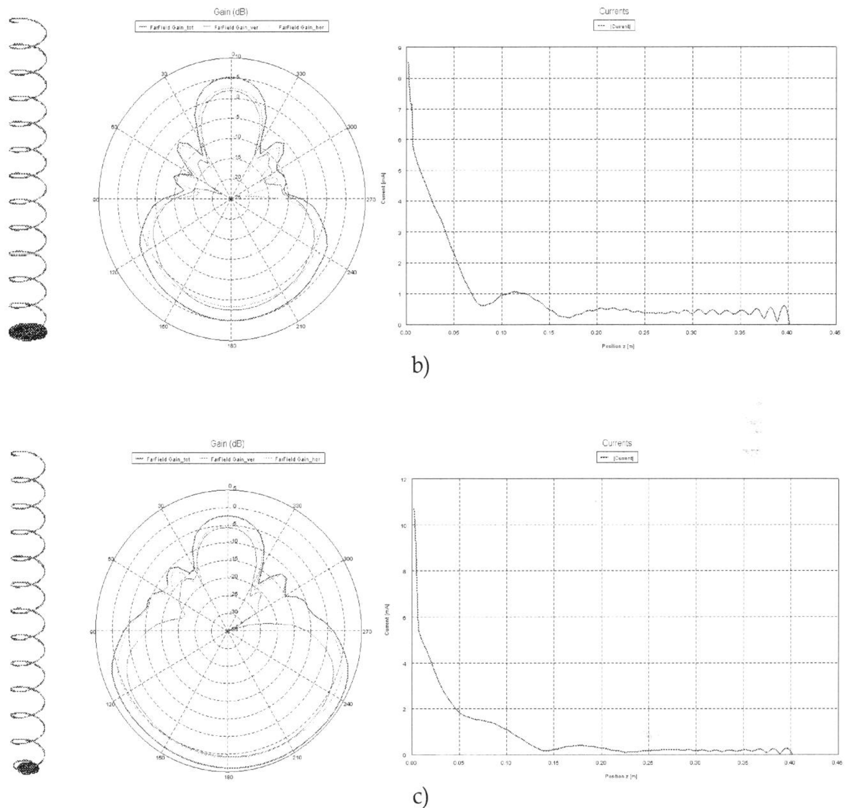

Fig. 10. The geometry, radiation pattern and current distribution of helical antenna with reflector of the diameter of a) $d_1 = 0.7\lambda$, b) $d_2 = 0.35\lambda$, and c) $d_3 = 0.3\lambda$.

3. Multifilar helical antennas

Beside the parameter modifications of monofilar helical antenna, the multiple number of wires in helix structure also offers interesting radiation performances for satellite communications. While monofilar helices are usually employed in transmission (Kraus, 1988), the multifilar helical antennas, bifilar and quadrifilar are mostly utilized at reception where wide beamwitdh coverage is needed to track as many of the visible satellites as possible (Kilgus, 1974; Lan et al., 2004).

3.1 The bifilar helical antenna

Patton was the first to describe bifilar helical antenna (BHA) with backfire radiation achieving maximum directivity just above the cut-off frequency of the main mode of the helical waveguide. The beamwidth broadens with frequency and for pitch angles of about forty five degrees, the beam splits and turns into a scanning mode toward broadside direction. As opposed to monofilar helical antenna, the backfire BHA radiates toward the feed point, its gain is independent of length (provided that the length is large enough) and the beamwidth increases with frequency (Patton, 1962).

Backfire bifilar helix is often used as a feed antenna because of its high efficiency, circularly polarized backward wave and low aperture blockage. In mobile handsets and various aerodynamic surfaces requiring low profile antennas side fed bifilar helical antenna can be used which produces a slant 45° linearly polarized omnidirectional toroidal pattern providing higher diversity gain in all directions (Amin et al., 2007).

In order for the bifilar helix to operate as backfire antenna, it is necessary that the currents flowing from the terminals to the ends of two helices are out of phase and the currents in the reversed direction are in phase. Hence, no radiation in forward direction is possible. This could be explained by the nature of the backward wave of current, where the phase is progressing toward the feed and the group velocity must be away from the feed point. A ground plane is not necessary in bifilar helical antenna design but this antenna usually achieves poor front-to-back (F/B) ratio which can cause interference problems when used as a receiving antenna. However, bifilar helical antenna with tapered feed end improves F/B ratio as well as the antenna power gain and axial ratio in comparison with conical and standard bifilar helical antenna (Yamauchi et al., 1981).

The BHA simulations are carried out in FEKO software on the basis of the following parameters (Yamauchi et al., 1981); the wavelength λ = 10 cm, circumference of the helical cylinder C $=\lambda$, the pitch angle ψ = 12.5°, wire radius r = 0.005λ, tapering cone angle θ = 12.5° and the number of turns in tapered section n_t = 2.3 and in uniform section n_u = 3. Three types of BHA with the same axial length were simulated: standard, conical and tapered BHA, Fig. 11 a). Tapered BHA is consisted of two sections of equal axial lengths, one corresponding to the first half of the conical BHA and the other to the half of the standard BHA. According to the radiation patterns in Fig. 11 b) and the results given in the Table 1, the tapered BHA provides the best performance of the BHA considering the F/B ratio and gain with satisfying axial ratio and decreased HPBW. It is important to note that the conical and tapered BHA's give better radiation characteristics than the standard BHA. Further investigation of the tapered BHA in terms of height reduction concerning the growing need for antenna miniaturization, shows that good BHA performance can be achieved with even smaller tapered bifilar helical antenna. The height of this antenna was reduced with a step of one spacing of the standard BHA (p = C tanψ) and the results are summarized in Table 2. The simulations obtained for the reduced version of tapered BHA yielded the best results for the one with n_u = 1 and n_t = 2.3 which corresponds to 2/3 of the total length of the original BHA, with the geometry and radiation pattern shown in Fig. 12.

In order to reduce the antenna length, Nakano et al. examined bifilar scanning helical antenna with large pitch angle terminated with a resistive load. This antenna generates circularly polarized scanning radiation pattern from backfire to normal. The simulations show the scanning radiation patterns of the bifilar helix with six turns, pitch angle of 68° and diameter of 1.6 cm, through the frequency band from 1.3 – 2.5 GHz (Nakano et. al, 1991). Fig. 13 illustrates typical radiation patterns, the backfire conical and normal radiation pattern reaching the antenna gain of 10 dB, Fig. 13 a) and b), respectively.

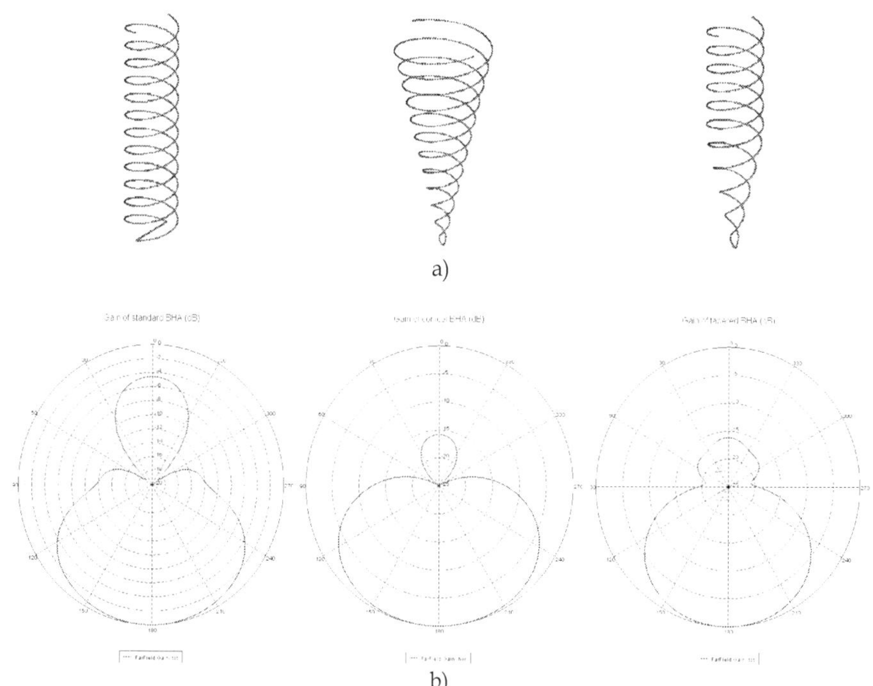

a)

b)

Fig. 11. a) Standard, conical and tapered BHAs, and b) their radiation patterns.

	F/B (dB)	Gain (dB)	AR	HPBW (°)
Standard BHA	4.5	5.6	0.79	111
Conical BHA	15.6	6.5	0.92	113
Tapered BHA	16	7.6	0.76	87

Table 1. Simulation results of radiation characteristics of standard, conical and tapered BHA.

	F/B (dB)	Gain (dB)	AR	HPBW (°)
Tapered BHA ($n_t = 1.5$, $n_u = 3$)	15.4	7.1	0.72	90
Tapered BHA ($n_t = 0.8$, $n_u = 3$)	11.2	5.7	0.89	120
Tapered BHA ($n_u = 0$, $n_t = 2.3$)	7.5	6	0.72	85
Tapered BHA ($n_u = 1$, $n_t = 2.3$)	14.8	7.8	0.65	82
Tapered BHA ($n_u = 2$, $n_t = 2.3$)	14.0	7.8	0.75	87

Table 2. Simulation results of reduced size tapered BHA.

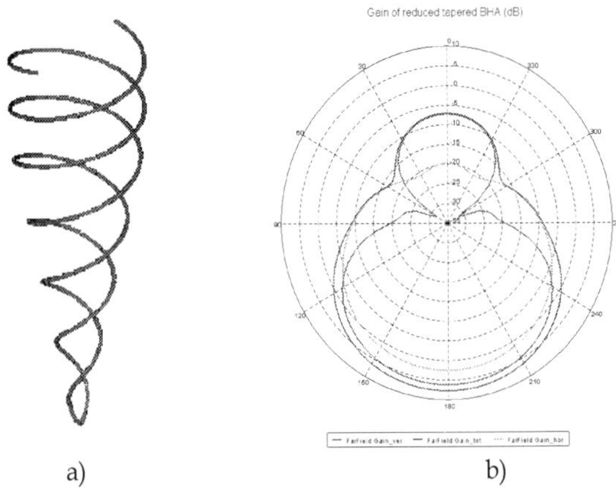

a)   b)

Fig. 12. Geometry and radiation patterns of reduced size BHA, a) and b) respectively.

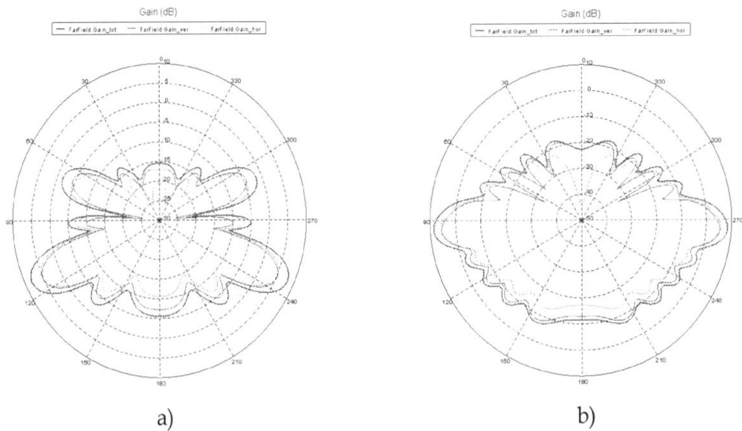

a)   b)

Fig. 13. Typical radiation patterns of bifilar scanning helical antenna, a) conical at 1.6 GHz
and b) normal radiation pattern at 2.1 GHz.

Contrary to monofilar helical antenna, the bifilar helical antenna yields scanning radiation
mode when relative phase velocity $p = v/c = 1.0$. This is confirmed with the comparison
of the simulated results with the experimental and calculated results (Nakano et al.,
1991; Zimmerman, 2000) of the lobe direction for the different values of phase velocity,
Fig. 14.

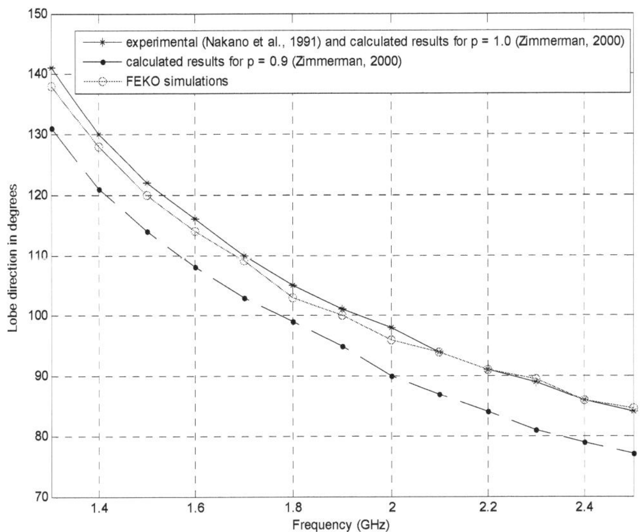

Fig. 14. The comparison of the simulated, calculated and experimental results for the lobe direction vs. frequency.

3.2 The quadrifilar helical antenna

The quadrifilar helical antenna (QHA), also known as the Kilgus coil, is mostly used for telemetry, tracking and command (TT&C) satellite systems due to its simplicity, small size, wide circularly polarized beam and insensitivity to nearby metal objects. The QHA consists of four helical wires equally spaced circumferentially and fed from the top or the bottom. The open ended QHA generally uses the length of each wire of $\lambda/4$ or $3\lambda/4$ with typical input impedance in the range 10 to 20 ohms while the short–circuited QHA uses $\lambda/2$ or λ length of each wire which produces resonant input impedance of nearly 50 ohms. Printed QHAs, convenient for high frequency applications, are manufactured using the dielectric substrate (Chew et al., 2002; Hanane et al., 2007) while wire QHA-s can be implemented on cylindrical, conical, square and spherical dielectric mechanical supports (Casey & Bansal, 2002; Hui et al., 2001). The size reduction of quadrifilar helical antennas can be achieved with geometrical reduction techniques such as sinusoidal (Fonseca et al., 2009; Takacs et al., 2010), rectangular (Ibambe et al., 2007), meander line (Chew et al., 2002) and other techniques (Letestu et al., 2006).

Radiation pattern of fractional turn resonant QHA is cardioid-shaped and circularly polarized with wide beamwitdh, but by extending the fractional-turn QHA to an integral number of turns shaped-conical radiation pattern can be obtained for many applications in spacecraft communications (Kilgus, 1975).

The Kilgus coil consisted of four wires $\lambda/2$ long and forming a ½ turn of a helix, generates a cardioid-shaped backfire radiation pattern with circular polarization and a very high HPBW when two pairs are fed in phase quadrature and lower ends are short-circuited (Kilgus, 1968, 1974). The antenna is fed with a split sheath balun and the phase quadrature is achieved by adjusting the lengths of the wires.

The performance of the QHA is described with the following parameters: the length of one element consisted of two radials and a helical section l_{el} (integer number of $\lambda/2$), axial length between the radials l_{ax} and the number of turns N. We designed a half turn QHA for GPS L2 signal with the central frequency of f = 1220 MHz and the following parameters: l_{el} = $\lambda/2$, wire diameter d = 2 mm, bending radius b_r = 5 mm and width-to height ratio w/h = 0.44 (the length of wires was adjusted to achieve phase quadrature so width w is the longitudinal width and h is axial height (l_{ax}) of the antenna). This is the so called self-phased QHA where the wire of one bifilar helix is longer than the resonant length, so that the current has a phase lead of 45° and the other is shorter in order to achieve a phase lag of 45°. Instead of infinite balun, we proposed a stripline structure for impedance matching and the support for helical wire. Fig. 15 c) shows that matching stripline is made of shorter part designed to counteract the imaginary part of the antenna input impedance and longer quarterwave part which is used to tune the real component of antenna input impedance to 50-Ω coaxial line impedance (Sekelja et al., 2009).

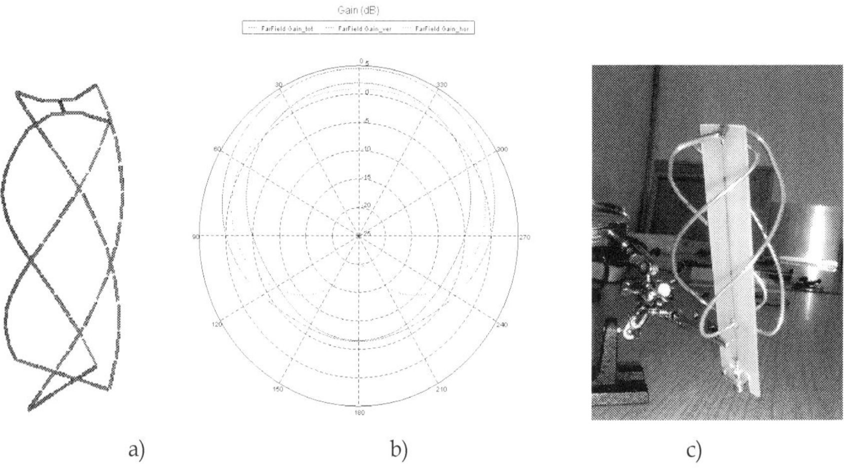

a)   b)   c)

Fig. 15. The geometry with wire segments a) and simulated radiation patterns b) of QHA and c) the antenna prototype with stripline feeding structure.

In many satellite applications, it is also desirable to concentrate the radiated energy into a shaped conical beam with full cone angles from 120° to 180° (Kilgus, 1975). So, for the same frequency, f = 1220 MHz, we simulated a three turn QHA (Fig. 16 a)) fed in phase quadrature with short circuited ends which achieves gain decreasing from the maximum of 5.6 dB at the edge of the cone to the local minimum of -2.5 dB at the centre. Radiation pattern in Fig. 16 b) also shows that this antenna gives an excellent axial ratio.

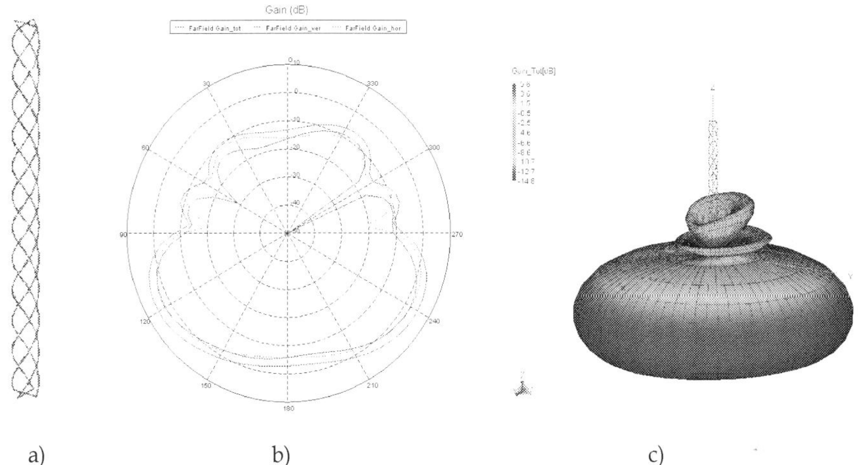

a)   b)   c)

Fig. 16. a) The geometry, b) the 2D and c) 3D simulated radiation patterns of three turn QHA.

5. Conclusion

In this chapter, the basic theory and simulations of helical antennas are presented. It is shown that various radiation patterns can be obtained with conventional helical antenna and its modifications: forward and backward radiation, beam, normal and scanning radiation, from hemispherical to conical-shaped radiation patterns. The circular polarization is easily achieved (except for the normal mode) and it can be improved by end tapering. These modifications include the change of helix geometry, the size and shape of reflector, the number of wires and implementing some guiding structure.

However, when implementing real materials in practical design, they must be evaluated for their influence on the overall antenna performance. Thus, while the depicted analytical approach offers a tool for the optimal design and basic analysis of the helical antenna, although not completely impossible, it becomes too complex to be implemented in final decision about the practical design. The performances of the designed antenna must therefore be tested by some numerical tool or by measurements.

6. References

Adekola, A. S., Mowete, A. I. & Ayorinde, A. A. (2009). Compact theory of the broadband elliptical helical antenna, *European Journal of Scientific research*, Vol. 31, No. 3, (2009), pp. 446-490, ISSN 1450-216X

Amin, M., Cahill, R. & Fusco, V. Single feed low profile omnidirectional antenna with slant 45° linear polarization, *IEEE Transactions on Antennas and Propagation*, Vol. 55, No. 11, (November 2007), pp. 3087-3090, ISSN 0018-926X

Barts, R. M. & Stutzman, W. L. (1997). A reduced size helical antenna, *Proceedings of IEEE Antennas and Propagation Society International Symposium*, ISBN 0-7803-4178-3, Montreal, Canada, July 1997.

Blazevic, Z. & Skiljo, M. (2010). Bandwidth of the Helical Beam Antenna Loaded by a Dielectric Rod, *Proceedings of ICECom*, ISBN 978-1-61284-998-0, Dubrovnik, Croatia, September 2010.

Bulgakov, B. M., Shestopalov, V. P., Shiskin, L. A. & Yakimenko, I. P. (1960). Symmetrical surface waves in a helix waveguide with a ferrite medium, *Journal of Radio and Electronic Physics*, Vol. 5, No. 11, (1960), pp. 102-119

Carver, K. R. (1967). The helicone-a circularly polarized antenna with low side-lobe level, *Proceedings of the IEEE*, Vol. 55, No. 4, (April 1967), p. 559, ISSN 0018-9219

Casey, J. P. & Basal, R. (1988). Dielectrically loaded wire antennas, *Proceedings of the IEEE*, Vol. 135, No. 2, (April 1988), pp. 103-110, ISSN 0950-107X

Casey, J. P. & Basal, R. (1988) Square helical antenna with a dielectric core, *IEEE Transactions on Electromagnetic Compatibility*, Vol. 30, No. 4, (November 1988), pp. 429-436, ISSN 0018-9375

Cha, A. G. (1972). Wave propagation on helical antennas, *IEEE Transactions on Antennas and Propagation*, Vol. 20, No. 5, (September 1972), pp. 556-560, ISSN 0018-926X

Chew, D. K. C. & Saunders, S. R. (2002). Meander line technique for size reduction of quadrifilar helix antenna, *IEEE Antennas and Wireless Propagation Letters*, Vol. 1, No. 1, (2002.) pp. 109-111, ISSN 1536-1225

Djordjevic, A. R., Zajic, A. G. & Ilic, M. M. (2006). Enhancing the gain of helical antennas by shaping the ground conductor, *IEEE Antennas and Wireless Propagation Letters*, Vol. 5, No. 1, (December 2006), pp. 138-140, ISSN 1536-1225

Fonseca, N. J. G. & Aubert, H. (2009). Very compact quadrifilar helix antenna in VHF band with quasi hemispherical radiation pattern, *Proceedings of IEEE Antennas and Propagation Society International Symposium*, ISBN 978-1-4244-3647-7, Charleston, South Carolina, June 2009.

Hanane, L., Hebib, S., Aubert, H. & Fonseca, N. J. G. (2007). Compact printed quadrifilar helix antennas for stratospheric ballons telemetry, *Proceedings of IEEE Antennas and Propagation Society International Symposium*, ISBN 978-1-4244-0877-1, Honolulu, Hawai, June 2007.

Hui, H. T., Yung, E. K. N. & Leung, K. W. (1997). Numerical and experimental studies of a helical antenna loaded by a dielectric resonator, *Radio Science*, Vol. 32, No. 2, (March-April 1997), pp. 295-304, ISSN 0048-6604

Hui, H. T., Chan, K. Y. & Yung, E. K. N. (2001). The input impedance and the antenna gain of the spherical helical antenna, *IEEE Transactions on Antennas and Propagation*, Vol. 49, No. 8, (August 2001), pp. 1235-1237, ISSN 0018-926X

Ibambe, M. G., Letestu, Y. & Sharaiha, A. (2007). Compact printed quadrifilar helical antenna. Electronic Letters, Vol. 43, No. 13, (June 2007), pp. 697-698, ISSN 0013-5194

Kilgus, C. (1968). Multielement, Fractional Turn Helices. *IEEE Transactions on Antennas and Propagation*, Vol.16, No.4, (July 1968), pp. 499-500 , ISSN 0018-926X

Kilgus, C. (1974). Spacecraft and Ground Station Applications of the Resonant Quadrifilar Helix, *Proceedings of IEEE Antennas and Propagation Society International Symposium*, Vol.12, pp. 75-77, June 1974.

Kilgus, C. (1975). Shaped-Conical Radiation Pattern Performance of the Backfire Quadrifilar Helix. *IEEE Transactions on Antennas and Propagation*, Vol. 23, No. 3, (May 1975), pp. 392-397 , ISSN 0018-926X

Klock, P. (1963). A study of wave propagation of helices, University of Illinois Antenna Laboratory Technical Report No. 68, March 1963.

Kraft, U. R. & Mönich, G. (1990). Main-beam polarization properties of modified helical antennas, *IEEE Transactions on Antennas and Propagation*, Vol. 38, No. 5, (May 1990), pp. 589-597, ISSN 0950-107X

Kraus, J. D. & Williamson J. C. (1948). Characteristic of helical antennas radiating in the axial mode, *Journal of Applied Physics*, Vol. 19, No. 1, (January 1948), pp. 87-96, ISSN 0021-8979

Kraus, J. D. (1949). The helical antenna, *Proceedings of IRE*, Vol. 37, No. 3, (March 1949), pp. 263-272, ISSN 0096-8390

Kraus, J. D. (1988). *Antennas (2nd ed)*, McGraw-Hill Companies, ISBN 978-0070354227, New Delhi, India

Lan, C. W., Chang, T. H. & Kiang, J. F. (2004). Helical antenna for GPS applications, *Proceedings of IEEE Antennas and Propagation Society International Symposium*, ISBN 0-7803-8302-8, June 2004.

Letestu, Y. & Sharaiha, A. (2006). Broadband folded printed quadrifilar helical antenna, *IEEE Transactions on Antennas and Propagation*, Vol. 54, No. 5, (May 2006), pp. 1600-1604, ISSN 0018-926X

Maclean, T. S. M. & Kouyoumjian, R. G. (1959). The bandwidth of helical antennas, *IRE Transactions on Antennas and Propagation*, Vol. 7, No. 5, (December 1959), pp. 379-386, ISSN 0096-1973

Marsh, J. (1950). Current distributions on helical antennas, Project Report No. 339-10, The Ohio State University Research Foundation, February 28, 1950.

Mimaki, H. & Nakano, H. (1998). Double pitch helical antenna, *Proceedings of IEEE Antennas and Propagation Society International Symposium*, ISBN 0-7803-4478-2, Atlanta, Georgia, June 1998.

Nakano, H., Samada, Y. & Yamauchi, J. (1986). Axial mode helical antenna, *IEEE Transactions on Antennas and Propagation*, Vol. AP-34. No. 9, (September 1986), pp. 1143-1148, ISSN 0018-926X

Nakano, H., Yamauchi, J. & Mimaki, H. (1988). Backfire radiation of a monofilar helix with a small ground plane, *IEEE Transactions on Antennas and Propagation*, Vol. 36, No. 10, (October 1988.), pp. 1359-1364, ISSN 0018-926X

Nakano, H., Takeda, H., Honma, T., Mimaki, H. & Yamauchi, J. (1991.). Extremely low-profile helix radiating a circularly polarized wave, *IEEE Transactions on Antennas and Propagation*, Vol. 39, No. 6, (June 1991), pp. 754-757, ISSN 0018-926X

Nakano, H., Mimaki, H. & Yamauchi, J. (1991). Loaded bifilar helical antenna with small radius and large pitch angle, *Electronic Letters*, Vol. 27, No. 17, (August 1991), pp. 1568–1569, ISSN 0013-5194

Neureuther, A. R., Clock, P. W. & Mittra, R. (1967). A study of the sheath helix with a conducting core and its application to the helical antenna, *IEEE Transactions on Antennas and Propagation*, Vol. AP-15, No. 2, (March 1967), pp. 203-210, ISSN 0018-926X

Olcan, D. I., Zajic, A. R., Ilic, M. M. & Djordjevic, A.R. (2006). On the optimal dimensions of helical antenna with truncated-cone reflector, *Proceedings of EuCAP*, ISBN 978-92-9092-937-6, Nice, France, November 2006.

Patton, W. T. (1962). The backfire bifilar helical antenna, Technical Report No. 61, Electrical Engineering Research Laboratory, University of Illionois, September 1962.

Sekelja, M. , Jurica, J. & Blazevic, Z. (2009). Designing and testing the quadrafilar helical antenna, *Proceedings of SoftCOM*, ISBN 978-1-4244-4973-6, Hvar, Croatia, September 2009.

Skiljo, M. , Blazevic, Z., Jurisic, A. and Pandzic, K. (2010). Improving the Helical Antenna Performance by Changing the Pitch Angle and the Shape of Reflector, *Proceedings of SoftCOM*, ISBN 978-1-4244-4973-6, Bol, Croatia, September 2010.

Sensiper, S. (1951). Electromagnetic wave propagation on helical conductors, Technical Report No. 194, MIT Research Laboratory of Electronic, May 16, 1951.

Sensiper, S. (1955). Electromagnetic wave propagation on helical structures, *Proceedings of IRE*, ISSN 0096-8390, February 1955.

Shestopalov, V. P., Bulgakov, A. A. & Bulgakov, B. M. (1961). Theoretical and experimental studies of helical dielectric antennas, *Journal of Radio and Electronic Phys*ics, Vol. 6, July 1961, pp. 1011-1019.

Sultan, N., Moody, M., Whelpton, J. & Hodgson, C. (1984). Novel broadband double pitch cylindrical helical antenna for satellite and ground applications, *Proceedings of IEEE Antennas and Propagation Society International Symposium*, Vol.22, pp. 162-165, June 1984.

Takacs, A., Fonseca, N. J. G. & Aubert, H. (2010). Height reduction of the axial-mode open-ended quadrifilar helical antenna, *IEEE Antennas and Wireless Propagation Letters*, Vol.9, (September 2010.) pp. 942-945 , ISSN 1536-1225

Vaughan, R. G. & Andersen, J. B. (1985). Polarization properties of the axial mode helix antenna, *IEEE Transactions on Antennas and Propagation*, Vol 33, No. 1, (January 1985), pp. 10-20, ISSN 0018-926X

Wong, J. L. & and King, H. E. (1979). Broadband quasi-taper helical antennas, *IEEE Transactions on Antennas and Propagation*, Vol. 27, No. 1, (January 1979), pp. 72-78, ISSN 0018-926X

Wong, J. L. & and King, H. E. (1982). Empirical Helix Antenna Design, *Proceedings of IEEE International Symposium on Antennas and Propagation*, p.p. 366-369, May 1982.

Yamauchi, J. , Nakano, H. & Mimaki, H. (1981). Backfire bifilar helical antenna with tapered feed end, *Proceedings of IEEE Antennas and Propagation Society International Symposium*, Vol. 19, pp. 683-686, June 1981.

Zimmerman, R. K., Jr. (2000). Traveling wave analysis of a bifilar scanning helical antenna, *IEEE Transactions on Antennas and Propagation*, Vol 48, No. 6, (June 2000), pp. 1007-1009, ISSN 0018-926X

Part 2

Atmospheric Effects in Satellite Links over Ka Band

Theoretical Analysis of Effects of Atmospheric Turbulence on Bit Error Rate for Satellite Communications in Ka-band

Tatsuyuki Hanada[1], Kiyotaka Fujisaki[2] and Mitsuo Tateiba[3]
[1]*Japan Aerospace Exploration Agency*
[2]*Kyushu University*
[3]*Ariake National College of Technology*
Japan

1. Introduction

In electromagnetic wave propagation through the earth's atmosphere like satellite communications, it is known that random fluctuations of the dielectric constant of atmosphere affect propagation characteristics of electromagnetic waves (Fante, 1975; 1980; Ishimaru, 1997; Rytov et al., 1989; Strohbehn, 1977; Tatarskii, 1961; 1971; Tatarskii et al., 1993; Uscinski, 1977; Wheelon, 2003). The random fluctuations, called atmospheric turbulence, cause spot dancing, wave form distortion, scintillations of the received intensity, the decrease in the spatial coherence of wave beams etc. These effects make the received power decrease, and result in the degradation in the performance on satellite communication links. Fig. 1 shows the image of spot dancing and wave form distortion of wave beams. Fig. 2 presents the image of the decrease in the spatial coherence of transmitted waves due to a wave front distortion.

The effects of atmospheric turbulence are not negligible in satellite communications in high carrier frequencies at low elevation angles. For example, tropospheric scintillation, caused by turbulence in the lowest layer of atmosphere, has been observed in satellite communications in Ku-band at low elevation angles (Karasawa, Yamada & Allnutt, 1988; Karasawa, Yasukawa & Yamada, 1988). Therefore, it becomes important to consider the effects of atmospheric turbulence appropriately in the design of such satellite communication systems. Some models to predict tropospheric scintillation have been developed for applications up to around 14 GHz in the carrier frequency on the basis of both theoretical and empirical studies (Ippolito, 2008). However, because a carrier frequency becomes higher according to the increase in the required channel capacity of satellite communication links in the next generation, the analysis of the effects of atmospheric turbulence should be done for applications at the higher carrier frequencies such as Ka-band, a millimeter wave and an optical wave. Some studies are conducted for satellite communications in such frequencies (Marzano et al., 1999; Matricciani et al., 1997; Matricciani & Riva, 2008; Mayer et al., 1997; Otung, 1996; Otung & Savvaris, 2003; Peeters et al., 1997).

We study the effects of atmospheric turbulence on satellite communications in such high frequencies by the theoretical analysis of the moments of wave fields given on the basis of a multiple scattering method (Tateiba, 1974; 1975; 1982). We investigate the method to estimate

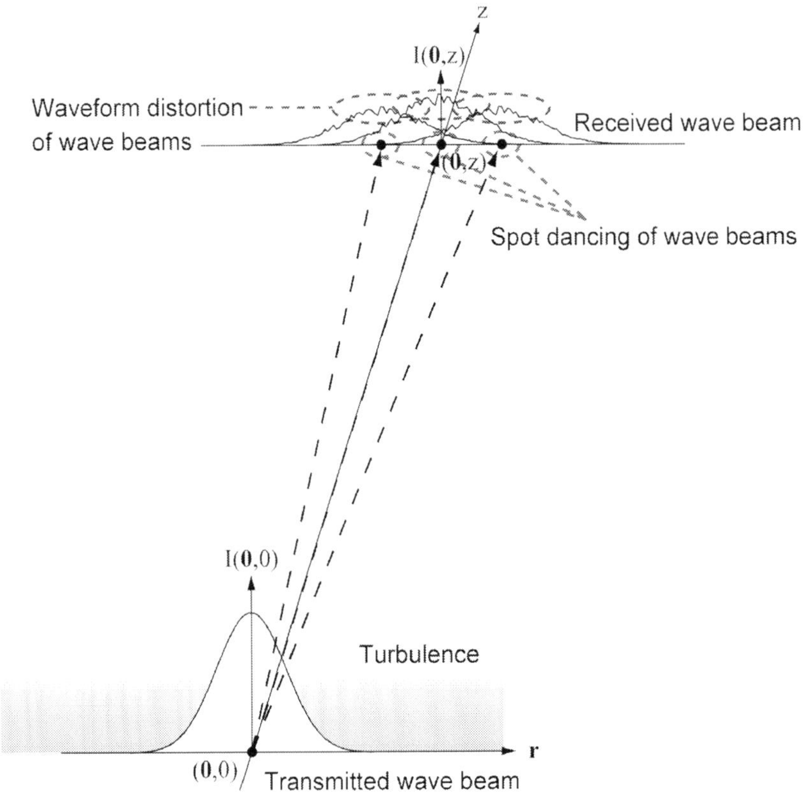

Fig. 1. Spot dancing and wave form distortion of wave beams through atmospheric turbulence where $I(\mathbf{r}, z)$ denotes the intensity of a wave beam at $(\mathbf{r}, z)$.

the effect on bit error rate (BER) which is one of the most important parameters to determine the system performance (Hanada et al., 2008a;b; 2009a;b;c;d). The probability density function (PDF) of E_b/N_0 (the energy per bit to noise power density ratio) is needed in the analysis of BER for satellite communications. However, it is very difficult to derive the arbitrary order moment and the PDF by the multiple scattering theory, so that the alternative method to estimate effects of atmospheric turbulence on BER has to be considered.

In this chapter, we give attention to the average value of received power which can be obtained by the second moment of a Gaussian wave beam, and then we formulate BER derived from the average received power. We provide the method to estimate effects of atmospheric turbulence on satellite communications by analyzing the degradation in BER performance due to the decrease in the average received power.

Sec. 2 presents formulations which are used in the analysis of BER on satellite communications. We introduce the second moment of a Gaussian wave beam obtained by the moment equation. Using the second moment of a Gaussian wave beam, we prepare the

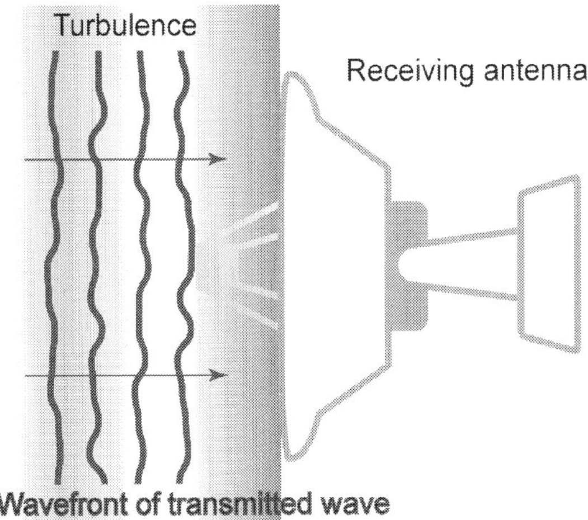

Fig. 2. Decrease in the spatial coherence of transmitted waves due to a wave front distortion.

modulus of the complex degree of coherence (DOC) and the BER derived from the average received power.

Sec. 3 shows the results of analysis of the effect of atmospheric turbulence on satellite communications in Ka-band at low elevation angles. We analyze the DOC and the BER derived from the average received power for the uplink and the downlink, respectively. Furthermore, we analyze the effect of atmospheric turbulence on the BER when we make an aperture radius of the ground station's antenna large in order to increase the antenna gain and improve BER performance.

Sec. 4 provides a summary of this chapter and subjects to resolve in future.

2. Formulation

2.1 Moment of received waves

We assume that an inhomogeneous random medium, which represents atmospheric turbulence, is characterized by fluctuations of the dielectric constant. The dielectric constant ε, the magnetic permeability μ and the conductivity σ are expressed as

$$\varepsilon = \varepsilon_0[1 + \delta\varepsilon(\mathbf{r}, z)] \tag{1}$$

$$\mu = \mu_0 \tag{2}$$

$$\sigma = 0, \tag{3}$$

where $\mathbf{r} = \mathbf{i}_x x + \mathbf{i}_y y$ ($\mathbf{i}_x$ and $\mathbf{i}_y$ denote the unit vectors of x and y coordinates, respectively) and ε_0 and μ_0 are the dielectric constant and the magnetic permeability for free space, respectively. The function $\delta\varepsilon(\mathbf{r}, z)$ is a Gaussian random function with the properties:

$$\langle \delta\varepsilon(\mathbf{r}, z) \rangle = 0 \tag{4}$$

$$\langle \delta\varepsilon(\mathbf{r}_1, z_1) \cdot \delta\varepsilon(\mathbf{r}_2, z_2) \rangle = B(\mathbf{r}_-, z_+, z_-), \tag{5}$$

where $\mathbf{r}_- = \mathbf{r}_1 - \mathbf{r}_2$, $z_+ = (z_1 + z_2)/2$, $z_- = z_1 - z_2$, $B(\mathbf{r}_-, z_+, z_-)$ is the correlation function of random dielectric constant and the bracket notation $\langle \cdot \rangle$ denotes an ensemble average of the quantity inside the brackets. Thus the medium fluctuates inhomogeneously in the z direction and homogeneously in the $\mathbf{r}$ direction. Moreover, we assume that for any z,

$$B(\mathbf{0}, z, 0) \ll 1 \tag{6}$$

$$kl(z) \gg 1, \tag{7}$$

where $k = 2\pi/\lambda$ is the wave number for free space and λ is the wave length. The wave length can be described by $\lambda = c/f$, where c and f are velocity of light and the carrier frequency, respectively. The function $l(z)$ is the local correlation length of $\delta\varepsilon(\mathbf{r}, z)$. The medium changes little the state of polarization of the wave under the conditions (6) and (7), and the present analysis can be made in the scalar approximation. In addition, the forward scattering and the small angle approximations can be applied.

We represent $u(\mathbf{r}, z)$ as a successively forward scattered wave with $\exp(-jwt)$ time dependence in the inhomogeneous random medium. Fig. 3 shows a model of wave propagation in the inhomogeneous random medium. An arbitrary order moment of $u(\mathbf{r}, z)$, which is defined as

$$M_{\mu\nu}(z) \equiv \left\langle \prod^{\mu} u(\mathbf{s}_m, z) \prod^{\nu} u^*(\mathbf{t}_n, z) \right\rangle, \tag{8}$$

satisfies the following moment equation (Tateiba, 1982):

$$\left[\frac{\partial}{\partial z} - j\frac{1}{2k} \left(\sum^{\mu} \nabla^2_{\mathbf{s}_m} - \sum^{\nu} \nabla^2_{\mathbf{t}_n} \right) - j(\mu - \nu)k \right] M_{\mu\nu}(z)$$

$$= -\left\{ \frac{k^2}{4} \int_0^z dz' \left[(\mu - \nu)^2 B\left(\mathbf{0}, z - \frac{z'}{2}, z' \right) + \sum_{m=1}^{\mu} \sum_{n=1}^{\nu} D\left(\mathbf{s}_m - \mathbf{t}_n, z - \frac{z'}{2}, z' \right) \right.\right.$$

$$\left.\left. - \sum_{m=1}^{\mu} \sum_{n>m}^{\mu} D\left(\mathbf{s}_m - \mathbf{s}_n, z - \frac{z'}{2}, z' \right) - \sum_{m=1}^{\nu} \sum_{n>m}^{\nu} D\left(\mathbf{t}_m - \mathbf{t}_n, z - \frac{z'}{2}, z' \right) \right] \right\} M_{\mu\nu}(z)$$

$$M_{\mu\nu}(0) = M_{\mu\nu}^{\text{in}}(0), \tag{9}$$

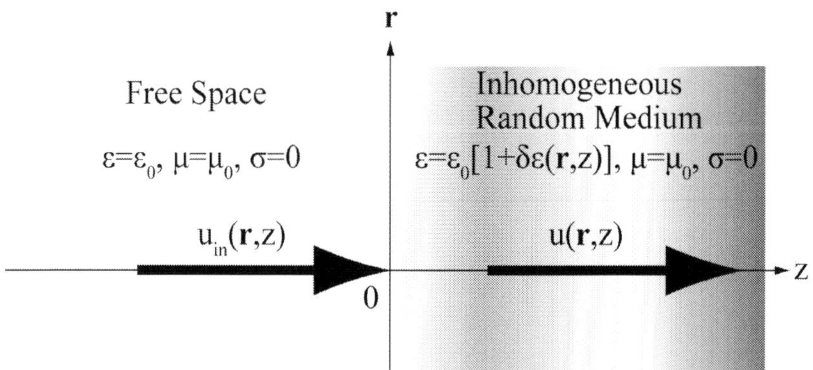

Fig. 3. Model of wave propagation in an inhomogeneous random medium.

where $\nabla = \mathbf{i}_x \partial/(\partial x) + \mathbf{i}_y \partial/(\partial y)$,

$$D\left(\mathbf{r}_-, z_+, z_-\right) = 2\left[B\left(\mathbf{0}, z_+, z_-\right) - B\left(\mathbf{r}_-, z_+, z_-\right)\right] \tag{10}$$

$$M_{\mu\nu}^{\text{in}}(z) = \prod^{\mu} u_{\text{in}}(\mathbf{s}_m, z) \prod^{\nu} u_{\text{in}}^*(\mathbf{t}_n, z), \tag{11}$$

and $u_{\text{in}}(\mathbf{r}, z)$ represents a transmitted waves which is a wave function in free space where $\delta\varepsilon(\mathbf{r}, z) = 0$. The function $D\left(\mathbf{r}_-, z_+, z_-\right)$ is the structure function of random dielectric constant. The exact solutions to (9), however, have not been obtained except for the second moment, which is one of the most important unsolved problems. The second moment of $u(\mathbf{r}, z)$ can be derived as follows (Tateiba, 1985):

$$M_{11}(\mathbf{r}_+, \mathbf{r}_-, z) = \langle u(\mathbf{r}_1, z)u^*(\mathbf{r}_2, z)\rangle$$

$$= \frac{1}{(2\pi)^2} \iint_{-\infty}^{\infty} d\boldsymbol{\kappa}_+ \, \hat{M}_{11}^{\text{in}}(\boldsymbol{\kappa}_+, \mathbf{r}_-, z)$$

$$\cdot \exp\left[j\boldsymbol{\kappa}_+ \cdot \mathbf{r}_- - \frac{k^2}{4} \int_0^z dz_1 \int_0^{z_1} dz_2 \, D\left(\mathbf{r}_- - \frac{z - z_1}{k}\boldsymbol{\kappa}_+, z_1 - \frac{z_2}{2}, z_2\right)\right], \tag{12}$$

where $\mathbf{r}_+ = (\mathbf{r}_1 + \mathbf{r}_2)/2$,

$$\hat{M}_{11}^{\text{in}}(\boldsymbol{\kappa}_+, \mathbf{r}_-, z) = \iint_{-\infty}^{\infty} d\mathbf{r}_+ \, M_{11}^{\text{in}}(\mathbf{r}_+, \mathbf{r}_-, z) \exp\left(-j\boldsymbol{\kappa}_+ \cdot \mathbf{r}_+\right) \tag{13}$$

$$M_{11}^{\text{in}}(\mathbf{r}_+, \mathbf{r}_-, z) = u_{\text{in}}(\mathbf{r}_1, z)u_{\text{in}}^*(\mathbf{r}_2, z). \tag{14}$$

2.2 Second moment for Gaussian wave beam

A transmitted wave is assumed to be a Gaussian wave beam, where the transmitting antenna is located in the plane $z = 0$ and the amplitude distribution is Gaussian with the minimum spot size w_0 at $z = -z_0$ and w_0 denotes the radius at which the field amplitude falls to $1/e$ of that on the beam axis (see Fig. 4). Then, the wave field in free space (Tateiba, 1985) is given by

$$u_{\text{in}}(\mathbf{r}, z) = \sqrt{\frac{2A}{\pi}} \frac{1}{w} \exp\left[-(1 - jp)\frac{r^2}{w^2} + j(kz - \beta)\right], \tag{15}$$

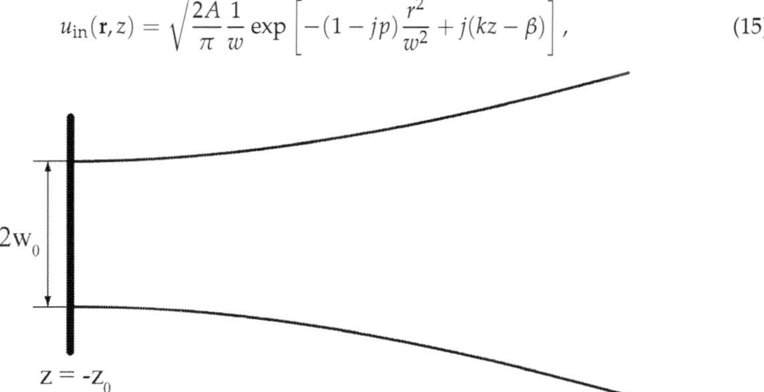

Fig. 4. Gaussian wave beam.

where A is constant, $r = |\mathbf{r}|$ and

$$w = w_0(1 + p^2)^{1/2} \tag{16}$$

$$p = 2(z + z_0)/(kw_0^2) \tag{17}$$

$$\beta = \tan^{-1} p. \tag{18}$$

Therefore,

$$M_{11}^{\text{in}}(\mathbf{r}_+, \mathbf{r}_-, z) = \frac{2A}{\pi w^2} \exp\left[-\frac{2}{w^2} r_+^2 + j\frac{2p}{w^2}(\mathbf{r}_+ \cdot \mathbf{r}_-) - \frac{r_-^2}{2w^2} \right] \tag{19}$$

$$\hat{M}_{11}^{\text{in}}(\boldsymbol{\kappa}_+, \mathbf{r}_-, z) = A \exp\left[-\frac{w^2}{8}\kappa_+^2 + \frac{p}{2}(\mathbf{r}_- \cdot \boldsymbol{\kappa}_+) - \frac{r_-^2}{2w_0^2} \right]. \tag{20}$$

Substituting (20) into (12), the second moment for a Gaussian wave beam is given by

$$M_{11}(\mathbf{r}_+, \mathbf{r}_-, z) = \frac{A}{(2\pi)^2} \iint_{-\infty}^{\infty} d\boldsymbol{\kappa}_+ \exp\left[-\frac{w^2}{8}\kappa_+^2 + \left(j\mathbf{r}_+ + \frac{p}{2}\mathbf{r}_- \right) \cdot \boldsymbol{\kappa}_+ - \frac{r_-^2}{2w_0^2} \right.$$
$$\left. -\frac{k^2}{4} \int_0^z dz_1 \int_0^{z_1} dz_2\, D\left(\mathbf{r}_- - \frac{z - z_1}{k}\boldsymbol{\kappa}_+, z_1 - \frac{z_2}{2}, z_2 \right) \right]. \tag{21}$$

2.3 Structure function of random dielectric constant

We assume that the correlation function of random dielectric constant defined by (5) satisfies the Kolmogorov model. We use the von Karman spectrum (Ishimaru, 1997) which is the modified model of the Kolmogorov spectrum to be applicable over all wave numbers κ for $|\mathbf{r}_- + \mathbf{i}_z z_-|$:

$$\Phi_n(\kappa, z_+) = 0.033C_n^2(z_+) \frac{\exp\left(-\kappa^2/\kappa_m^2\right)}{\left(\kappa^2 + 1/L_0^2\right)^{11/6}}, \quad 0 \leq \kappa < \infty \tag{22}$$

where $\kappa_m = 5.92/l_0$. Parameters $C_n^2(z_+)$, L_0 and l_0 denote the refractive index structure constant, the outer scale and the inner scale of turbulence, respectively.

Here, we assume that the dielectric constant is delta correlated in the direction of propagation, which is the Markov approximation (Tatarskii, 1971). On this assumption, $B(\mathbf{r}_-, z_+, z_-)$ can be expressed by using the Dirac delta function $\delta(z)$ as follows:

$$B(\mathbf{r}_-, z_+, z_-) = 16\pi^2\delta(z_-) \int_0^{\infty} d\kappa\, \kappa\Phi_n(\kappa, z_+)J_0(\kappa r_-), \tag{23}$$

where $J_0(z)$ is the Bessel function of the first kind and order zero. Therefore, we obtain the structure function defined by (10) as follows:

$$D(\mathbf{r}_-, z_+, z_-) = 32\pi^2\delta(z_-) \int_0^{\infty} d\kappa\, \kappa\Phi_n(\kappa, z_+) \left[1 - J_0(\kappa r_-)\right]$$
$$= \delta(z_-) \cdot \frac{96\pi^2}{5} \cdot 0.033C_n^2(z_+)L_0^{5/3} \left[1 - \Gamma\left(\frac{1}{6}\right)\left(\frac{r_-}{2L_0}\right)^{5/6} I_{-5/6}\left(\frac{r_-}{L_0}\right) \right.$$
$$\left. + \Gamma\left(\frac{1}{6}\right)\left(\frac{1}{\kappa_m L_0}\right)^{5/3} \sum_{n=0}^{\infty} \frac{1}{n!} \left(\frac{1}{\kappa_m L_0}\right)^{2n} {}_1F_1\left(-n - \frac{5}{6}; 1; -\frac{\kappa_m^2 r_-^2}{4}\right) \right], \tag{24}$$

where $\Gamma(z)$, $_1F_1(a;b;z)$ and $I_\nu(z)$ are the gamma function, the confluent hypergeometric function of the first kind and the modified Bessel function of the first kind, respectively. Note that we use the solution including an infinite series (Wang & Strohbehn, 1974) in order to ease the numerical analysis of the integral with respect to κ.

2.4 Model of analysis

We analyze effects of atmospheric turbulence on the GEO satellite communications for Ka-band at low elevation angles. Fig. 5 shows the propagation model between the earth and the GEO satellite. The earth radius, the altitude of satellite and the elevation angle are expressed by R, L and θ, respectively. A height of the top of atmospheric turbulence is shown by h_t. The z_L is the distance from a transmitting antenna to a receiving antenna:

$$z_L = \sqrt{(R+L)^2 - (R\cos\theta)^2} - R\sin\theta, \tag{25}$$

and z_{ht} is the distance of propagation through atmospheric turbulence:

$$z_{ht} = \sqrt{(R+h_t)^2 - (R\cos\theta)^2} - R\sin\theta. \tag{26}$$

Note that $z_L \gg z_{ht}$ is satisfied for the GEO satellite communications. Therefore, for the uplink, we can approximate $z - z_1 \simeq z$ in the integral with respect to z_1 in (21), and then express the second moment of received waves at the GEO satellite:

$$M_{11}(\mathbf{r}_+,\mathbf{r}_-,z_{\mathrm{UL}}) \simeq \frac{A}{(2\pi)^2} \iint_{-\infty}^{\infty} d\boldsymbol{\kappa}_+ \, \exp\left[-\frac{w^2}{8}\kappa_+^2 + \left(j\mathbf{r}_+ + \frac{p}{2}\mathbf{r}_-\right)\cdot\boldsymbol{\kappa}_+ - \frac{r_-^2}{2w_0^2} \right.$$
$$\left. -\frac{k^2}{4}\int_0^{z_{\mathrm{UL}}} dz_1 \int_0^{z_1} dz_2\, D\left(\mathbf{r}_- - \frac{z_{\mathrm{UL}}}{k}\boldsymbol{\kappa}_+, z_1 - \frac{z_2}{2}, z_2\right) \right], \tag{27}$$

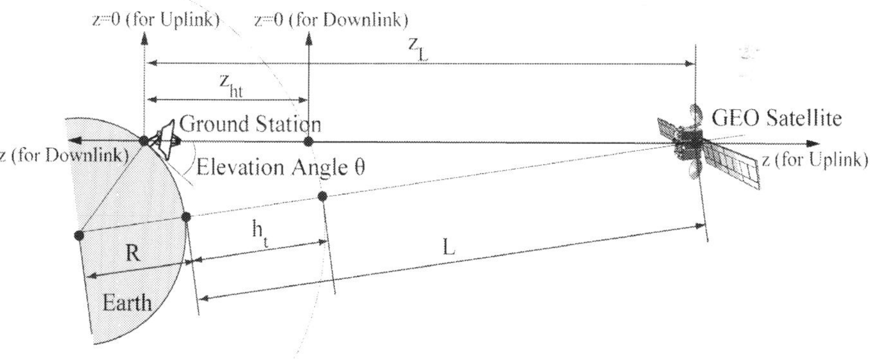

Fig. 5. Earth – GEO satellite propagation model.

where $z_{UL} = z_L$, $w = w_0\sqrt{1+p^2}$, $p = 2z_L/(kw_0^2)$ and the subscript of z_{UL} denotes the uplink. On the other hand, for the downlink, the statistical characteristics of a wave beam's incidence into atmospheric turbulence can be approximately treated as those of a plane wave's incidence. Thus the second moment of received waves at the ground station can be approximately expressed by

$$M_{11}(\mathbf{r}_+, \mathbf{r}_-, z_{DL}) \simeq \frac{2A}{\pi w^2} \exp\left[-\frac{k^2}{4}\int_0^{z_{DL}} dz_1 \int_0^{z_1} dz_2\, D\left(\mathbf{r}_-, z_1 - \frac{z_2}{2}, z_2\right)\right], \qquad (28)$$

where $z_{DL} = z_{ht}$, $w = w_0\sqrt{1+p^2}$, $p = 2(z_L - z_{h1})/(kw_0^2)$ and the subscript of z_{DL} denotes the downlink.

Here, the refractive index structure constant is assumed to be a function of altitude. Referring to some researches for the dependence of the refractive index structure constant in boundary layer (Tatarskii, 1971) and in free atmosphere (Martini et al., 2006; Vasseur, 1999) on altitude, we assume the following vertical profile as a function of altitude: $h = \sqrt{(z + R\sin\theta)^2 + (R\cos\theta)^2} - R$.

$$
\begin{aligned}
C_n^2(h) &= C_{n0}^2\left(1 + \frac{h}{h_{s1}}\right)^{-2/3}, & \text{for } 0 \le h < h_1 \\[2ex]
&= C_{n0}^2\left(1 + \frac{h_1}{h_{s1}}\right)^{-2/3}\left(\frac{h}{h_1}\right)^{-4/3}, & \text{for } h_1 \le h < h_2 \qquad (29) \\[2ex]
&= C_{n0}^2\left(1 + \frac{h_1}{h_{s1}}\right)^{-2/3}\left(\frac{h_2}{h_1}\right)^{-4/3}\exp\left(-\frac{h - h_2}{h_{s2}}\right), & \text{for } h_2 \le h \le h_t
\end{aligned}
$$

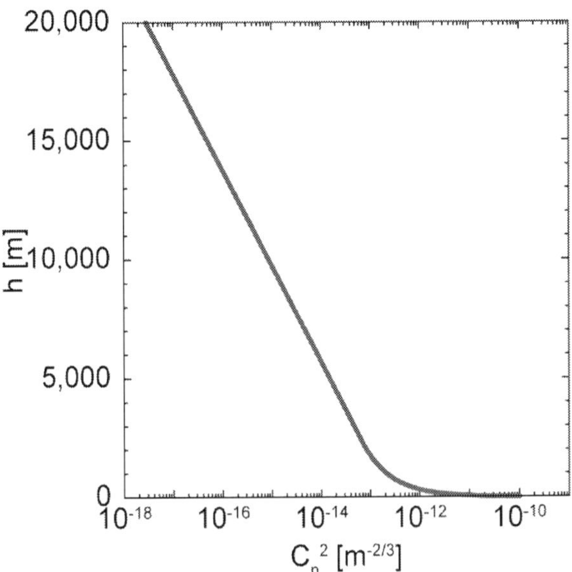

Fig. 6. Vertical profile of refractive index structure constant as a function of altitude.

ITEM	VALUE
Carrier frequency (uplink / downlink): f	30.0/20.0 GHz
Velocity of light: c	3.0×10^8 m/s
Elevation angle: θ	5.0 deg
Aperture radius of an antenna in the GEO satellite	1.2 m
Aperture radius of an antenna in the ground station	1.2 to 7.5 m
Earth radius: R	6,378 km
Height of GEO satellite: L	35,786 km
Height of the top of atmospheric turbulence: h_t	20 km
Refractive index structure constant at the ground level: C_{n0}^2	1.0×10^{-10} m$^{-2/3}$
Outer scale of turbulence: L_0	100 m
Inner scale of turbulence: l_0	1 mm

Table 1. Parameters used in analysis.

where $h_1 = 50$ m, $h_2 = 2,000$ m, $h_t = 20,000$ m, $h_{s1} = 2$ m and $h_{s2} = 1,750$ m. Fig. 6 shows a vertical profile of the refractive index structure constant. We assume $C_{n0}^2 = 1.0 \times 10^{-10}$ m$^{-2/3}$ by referring to the profile of the standard deviation value obtained by Reference (Vasseur, 1999). We set $L_0 = 100$ m and $l_0 = 1$ mm.
Table 1 shows parameters used in analysis.

2.5 Modulus of complex degree of coherence
Using the second moment of received waves, we examine the loss of spatial coherence of received waves on the aperture of a receiving antenna by the modulus of the complex degree of coherence (DOC) (Andrews & Phillips, 2005) defined by

$$\text{DOC}(\boldsymbol{\rho}, z) = \frac{M_{11}(\mathbf{0}, \boldsymbol{\rho}, z)}{[M_{11}(\boldsymbol{\rho}/2, \mathbf{0}, z) M_{11}(-\boldsymbol{\rho}/2, \mathbf{0}, z)]^{1/2}}, \tag{30}$$

where $\rho = |\boldsymbol{\rho}|$ is the separation distance between received wave fields at two points on the aperture as shown in Fig. 7.

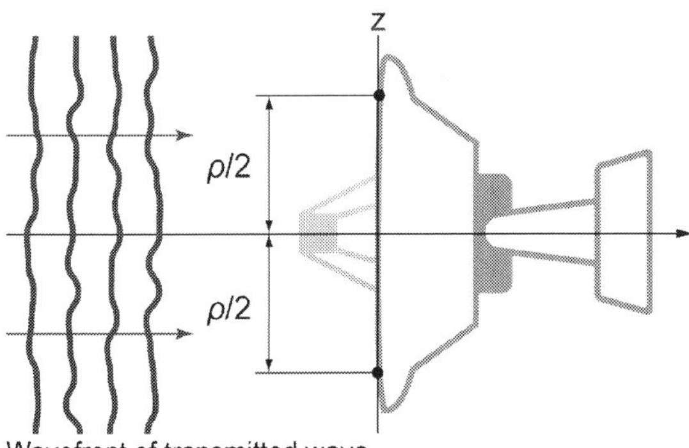

Fig. 7. Modulus of complex degree of coherence.

2.6 BER derived from average received power

We define BER derived from the average received power obtained by the second moment of received waves. Here we assume a parabolic antenna as a receiving antenna. When a point detector is placed at the focus of a parabolic concentrator, the instantaneous response in the receiving antenna is proportional to the electric field strength averaged over the area of the reflector. When the aperture size is large relative to the electromagnetic wavelength, the electric field strength averaged over the area of the reflector in free space can be described (Wheelon, 2003) by

$$\overline{u_{\text{in}}(z)} = \frac{1}{S_e} \iint_{-\infty}^{\infty} d\mathbf{r} \, u_{\text{in}}(\mathbf{r}, z) g(\mathbf{r}), \tag{31}$$

where S_e is the effective area of a reflector. The field distribution $g(\mathbf{r})$ is defined by a Gaussian distribution of attenuation across the aperture with an effective radius a_e:

$$g(\mathbf{r}) = \exp\left(-\frac{r^2}{a_e^2}\right). \tag{32}$$

Then the power received by the antenna in free space is given by

$$
\begin{aligned}
P_{\text{in}}(z) &= S_e \cdot \frac{\text{Re}\left[\overline{u_{\text{in}}(z)} \cdot \overline{u_{\text{in}}^*(z)}\right]}{Z_0} \\
&= \frac{1}{S_e Z_0} \cdot \text{Re}\left[\iint_{-\infty}^{\infty} \iint_{-\infty}^{\infty} d\mathbf{r}_+ d\mathbf{r}_- M_{11}^{\text{in}}(\mathbf{r}_+, \mathbf{r}_-, z) \exp\left(-\frac{2}{a_e^2} r_+^2 - \frac{1}{2a_e^2} r_-^2\right)\right],
\end{aligned} \tag{33}
$$

where $\text{Re}[x]$ denotes the real part of x and Z_0 is the characteristic impedance. The energy per bit E_b can be obtained by the product of the received power $P_{\text{in}}(z)$ and the bit time T_b:

$$
\begin{aligned}
E_b &= P_{\text{in}}(z) \cdot T_b \\
&= \frac{T_b}{S_e Z_0} \cdot \text{Re}\left[\iint_{-\infty}^{\infty} \iint_{-\infty}^{\infty} d\mathbf{r}_+ d\mathbf{r}_- \, M_{11}^{\text{in}}(\mathbf{r}_+, \mathbf{r}_-, z) \exp\left(-\frac{2}{a_e^2} r_+^2 - \frac{1}{2a_e^2} r_-^2\right)\right].
\end{aligned} \tag{34}
$$

We define the average energy per bit $\langle E_b \rangle$ affected by atmospheric turbulence as the product of the average received power and T_b. The average received power is given by the second moment of received waves:

$$\langle P(z) \rangle = \frac{1}{S_e Z_0} \cdot \text{Re}\left[\iint_{-\infty}^{\infty} \iint_{-\infty}^{\infty} d\mathbf{r}_+ d\mathbf{r}_- \, M_{11}(\mathbf{r}_+, \mathbf{r}_-, z) \exp\left(-\frac{2}{a_e^2} r_+^2 - \frac{1}{2a_e^2} r_-^2\right)\right]. \tag{35}$$

Therefore,

$$
\begin{aligned}
\langle E_b \rangle &= \langle P(z) \rangle \cdot T_b \\
&= \frac{T_b}{S_e Z_0} \cdot \text{Re}\left[\iint_{-\infty}^{\infty} \iint_{-\infty}^{\infty} d\mathbf{r}_+ d\mathbf{r}_- \, M_{11}(\mathbf{r}_+, \mathbf{r}_-, z) \exp\left(-\frac{2}{a_e^2} r_+^2 - \frac{1}{2a_e^2} r_-^2\right)\right].
\end{aligned} \tag{36}
$$

We consider QPSK modulation which is very popular among satellite communications. It is known that BER in QPSK modulation is defined by

$$\text{PE} = \frac{1}{2} \text{erfc}\left(\sqrt{\frac{E_b}{N_0}}\right), \tag{37}$$

where $\mathrm{erfc}(x)$ is the complementary error function. We define BER derived from the average received power in order to evaluate the influence of atmospheric turbulence as follows:

$$\mathrm{PE_P} = \frac{1}{2}\mathrm{erfc}\left(\sqrt{\frac{\langle E_b \rangle}{N_0}}\right). \tag{38}$$

And then, using E_b in free space obtained by (34), the BER can be expressed by

$$\mathrm{PE_P} = \frac{1}{2}\mathrm{erfc}\left(\sqrt{S_P \cdot \frac{E_b}{N_0}}\right), \tag{39}$$

where the normalized received power S_P is given by

$$S_P = \frac{\langle E_b \rangle}{E_b} = \frac{\langle P(z) \rangle}{P_{\mathrm{in}}(z)}$$

$$= \frac{\mathrm{Re}\left[\iint_{-\infty}^{\infty}\iint_{-\infty}^{\infty} d\mathbf{r}_+ d\mathbf{r}_- \, M_{11}(\mathbf{r}_+, \mathbf{r}_-, z) \exp\left(-\frac{2}{a_e^2}r_+^2 - \frac{1}{2a_e^2}r_-^2\right)\right]}{\mathrm{Re}\left[\iint_{-\infty}^{\infty}\iint_{-\infty}^{\infty} d\mathbf{r}_+ d\mathbf{r}_- \, M_{11}^{\mathrm{in}}(\mathbf{r}_+, \mathbf{r}_-, z) \exp\left(-\frac{2}{a_e^2}r_+^2 - \frac{1}{2a_e^2}r_-^2\right)\right]}. \tag{40}$$

If the DOC is almost unity where the decrease in the spatial coherence of received waves is negligible, the received power can be replaced with the integration of the intensity $I(\mathbf{r}, z) = |u(\mathbf{r}, z)|^2$ over the receiving antenna. The received intensity in free space $I_{\mathrm{in}}(z)$ and the average received intensity affected by atmospheric turbulence $\langle I(z) \rangle$ are respectively given by

$$I_{\mathrm{in}}(z) = \iint_{-\infty}^{\infty} d\mathbf{r} \, M_{11}^{\mathrm{in}}(\mathbf{r}, 0, z) \exp\left(-\frac{2r^2}{a_e^2}\right) \tag{41}$$

$$\langle I(z) \rangle = \iint_{-\infty}^{\infty} d\mathbf{r} \, M_{11}(\mathbf{r}, 0, z) \exp\left(-\frac{2r^2}{a_e^2}\right). \tag{42}$$

Under the condition where the DOC is almost unity, we can reduce the number of the surface integral in calculation of (40) and then obtain BER derived from the average received intensity as follows:

$$\mathrm{PE_I} = \frac{1}{2}\mathrm{erfc}\left(\sqrt{\frac{\langle E_b \rangle}{N_0}}\right) = \frac{1}{2}\mathrm{erfc}\left(\sqrt{S_I \cdot \frac{E_b}{N_0}}\right), \tag{43}$$

where the normalized average received intensity S_I is given by

$$S_I = \frac{\langle E_b \rangle}{E_b} = \frac{\langle I(z) \rangle}{I_{\mathrm{in}}(z)} = \frac{\iint_{-\infty}^{\infty} d\mathbf{r} \, M_{11}(\mathbf{r}, 0, z) \exp\left(-\frac{2r^2}{a_e^2}\right)}{\iint_{-\infty}^{\infty} d\mathbf{r} \, M_{11}^{\mathrm{in}}(\mathbf{r}, 0, z) \exp\left(-\frac{2r^2}{a_e^2}\right)}. \tag{44}$$

3. Results

3.1 Modulus of complex degree of coherence
3.1.1 Uplink

Substituting (24) and (27) into (30), the DOC at the GEO satellite in the uplink can be described by

$$
\mathrm{DOC}(\rho, z_{\mathrm{UL}}) = \int_0^\infty d\kappa_+ \int_0^{2\pi} d\theta\, \kappa_+ \exp\left[-\frac{w^2}{8}\kappa_+^2 + \frac{p}{2}\kappa_+\rho\cos\theta - \frac{\rho^2}{2w_0^2} \right.
$$

$$
\left. - H\left(\rho - \frac{z_L}{k}\kappa_+, 0, z_{\mathrm{ht}}\right)\right]
$$

$$
\cdot \left\{ 2\pi \int_0^\infty d\kappa_+\, \kappa_+ J_0\left(\frac{\kappa_+\rho}{2}\right) \exp\left[-\frac{w^2}{8}\kappa_+^2 - H\left(-\frac{z_L}{k}\kappa_+, 0, z_{\mathrm{ht}}\right)\right] \right\}^{-1}, \quad (45)
$$

where

$$
H\left(\rho', 0, z_{\mathrm{ht}}\right) = \frac{12}{5}(k\pi)^2 L_0^{5/3} \int_0^{z_{\mathrm{ht}}} dz_1\, 0.033 C_n^2(z_1)
$$

$$
\cdot \left[1 + \Gamma\left(\frac{1}{6}\right)\left(\frac{1}{\kappa_m L_0}\right)^{5/3} \sum_{i=0}^\infty \frac{1}{i!}\left(\frac{1}{\kappa_m L_0}\right)^{2i} {}_1F_1\left(-i - \frac{5}{6}; 1; -\frac{\kappa_m^2 \rho'^2}{4}\right)\right.
$$

$$
\left. - \exp\left(\frac{\rho'}{L_0}\right) {}_1F_1\left(-\frac{1}{3}; -\frac{2}{3}; -\frac{2\rho'}{L_0}\right)\right], \quad (46)
$$

and

$$
\rho' = |\rho'| = \begin{cases} \sqrt{\rho^2 - 2\rho\dfrac{z_L}{k}\kappa_+\cos\theta + \dfrac{z_L^2}{k^2}\kappa_+^2} & \text{in } \rho' = \rho - \dfrac{z_L}{k}\kappa_+ \\[2ex] \dfrac{z_L}{k}\kappa_+ & \text{in } \rho' = -\dfrac{z_L}{k}\kappa_+. \end{cases} \quad (47)
$$

Fig. 8 shows that the DOC in the uplink is almost unity within the size of an aperture diameter of the receiving antenna of the GEO satellite ($\rho \lesssim 2a_e$). It means that the spatial coherence of received waves keeps enough large within the receiving antenna.

3.1.2 Downlink

Substituting (24) and (28) into (30), the DOC at the ground station in the downlink is obtained by

$$
\mathrm{DOC}(\rho, z_{\mathrm{DL}}) = \exp\left[- H\left(\rho, z_L - z_{\mathrm{ht}}, z_L\right)\right], \quad (48)
$$

where

$$
H\left(\rho, z_L - z_{\mathrm{ht}}, z_L\right) = \frac{12}{5}(k\pi)^2 L_0^{5/3} \int_{z_L - z_{\mathrm{ht}}}^{z_L} dz_1\, 0.033 C_n^2(z_1)
$$

$$
\cdot \left[1 + \Gamma\left(\frac{1}{6}\right)\left(\frac{1}{\kappa_m L_0}\right)^{5/3} \sum_{i=0}^\infty \frac{1}{i!}\left(\frac{1}{\kappa_m L_0}\right)^{2i} {}_1F_1\left(-i - \frac{5}{6}; 1; -\frac{\kappa_m^2 r^2}{4}\right)\right.
$$

$$
\left. - \exp\left(\frac{r}{L_0}\right) {}_1F_1\left(-\frac{1}{3}; -\frac{2}{3}; -\frac{2r}{L_0}\right)\right]. \quad (49)
$$

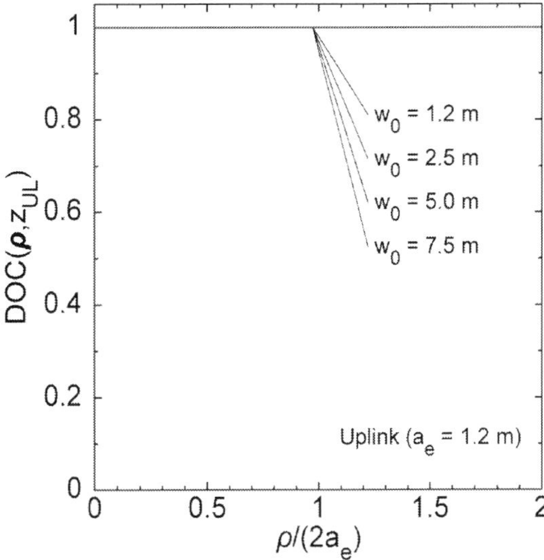

Fig. 8. The modulus of complex degree of coherence of received waves in the uplink for various beam radius at the transmitting antenna as a function of the separation distance between received wave fields at two points in the plane of the receiving antenna scaled by an aperture diameter of the receiving antenna $2a_e$.

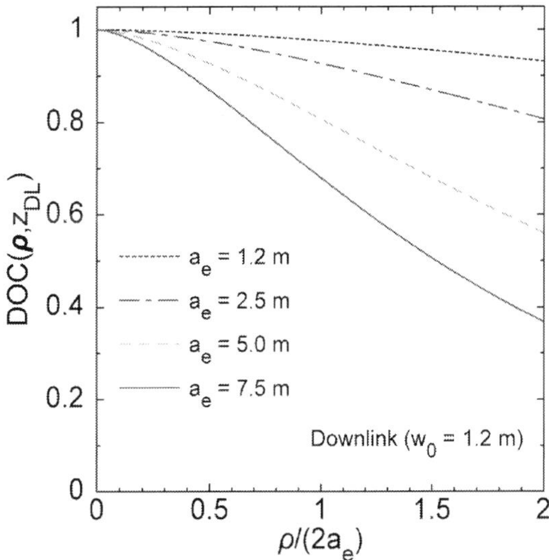

Fig. 9. Same as Fig. 8 except for the downlink where a beam radius at the transmitting antenna $w_0 = 1.2$ m for various aperture radius of the receiving antenna a_e.

As shown in Fig. 9, it is found that the decrease in the spatial coherence of received waves can not be neglected within a receiving antenna of the ground station. It indicates that an influence of the spatial coherence of received waves has to be considered in the analysis of BER in the downlink.

3.2 BER derived from average received power
3.2.1 Uplink

The BER derived from the average received intensity defined by (43) and (44) can be used for the uplink because the spatial coherence of received waves keeps enough large as shown in Fig. 8. Using (24), (27), (43) and (44), the BER can be expressed by

$$\mathrm{PE}_\mathrm{I} = \frac{1}{2}\mathrm{erfc}\left(\sqrt{S_I \cdot \frac{E_b}{N_0}}\right) \tag{50}$$

$$S_I = \frac{w^2 + a_e^2}{4}\int_0^\infty d\kappa_+ \, \kappa_+ \exp\left[-\frac{w^2 + a_e^2}{8}\kappa_+^2 - H\left(-\frac{z_L}{k}\kappa_+, 0, z_\mathrm{ht}\right)\right] \tag{51}$$

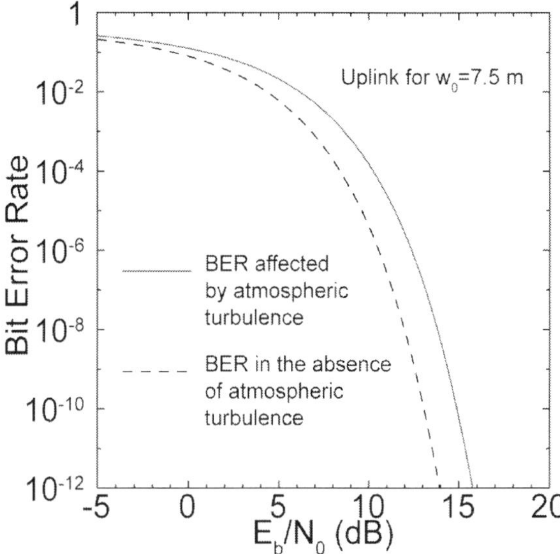

Fig. 10. BER derived from the average received intensity (PE_I) in the uplink in $w_0 = 7.5$ m.

$$H\left(-\frac{z_L}{k}\kappa_+,0,z_{\mathrm{ht}}\right) = \frac{12}{5}(k\pi)^2 L_0^{5/3}\int_0^{z_{\mathrm{ht}}} dz_1\, 0.033 C_n^2(z_1)$$

$$\cdot\left[1+\Gamma\left(\frac{1}{6}\right)\left(\frac{1}{\kappa_m^2 L_0^2}\right)^{5/6}\sum_{i=0}^{\infty}\frac{1}{i!}\left(\frac{1}{\kappa_m^2 L_0^2}\right)^i {}_1F_1\left(-i-\frac{5}{6};1;-\frac{\kappa_m^2 z_L^2}{4k^2}\kappa_+^2\right)\right.$$

$$\left.-\exp\left(\frac{z_L\kappa_+}{kL_0}\right){}_1F_1\left(-\frac{1}{3};-\frac{2}{3};-\frac{2z_L}{kL_0}\kappa_+\right)\right]. \tag{52}$$

Fig. 10 shows the BER affected by atmospheric turbulence in the uplink when wave beams are transmitted from the large aperture antenna where $w_0 = 7.5$ m. As reference, we plot a dashed line as the BER in the absence of atmospheric turbulence given by (37). It is found that BER increases compared with one in the absence of atmospheric turbulence. Because we have already shown that the decrease in the spatial coherence of received waves is negligible, we predict that the increase in BER is caused by the decrease in the average received intensity due to spot dancing shown in Fig. 1.

3.2.2 Downlink

For the downlink, the decrease in the spatial coherence of received waves can not be ignored as shown in Fig. 9. Therefore, we have to analyze the BER derived from the average received power defined by (39) and (40) which include an influence of the spatial coherence of received waves. Using (24), (28), (39) and (40), we obtain the BER as follows:

$$\mathrm{PE}_P = \frac{1}{2}\mathrm{erfc}\left(\sqrt{S_P\cdot\frac{E_b}{N_0}}\right) \tag{53}$$

$$S_P = \frac{1}{a_e^2}\int_0^{\infty} dr_-\, r_-\exp\left[-\frac{r_-^2}{2a_e^2}-H\left(\mathbf{r}_-,z_L-z_{\mathrm{ht}},z_L\right)\right] \tag{54}$$

$$H\left(\mathbf{r}_-,z_L-z_{\mathrm{ht}},z_L\right) = \frac{12}{5}(k\pi)^2 L_0^{5/3}\int_{z_L-z_{\mathrm{ht}}}^{z_L} dz_1\, 0.033 C_n^2(z_1)$$

$$\cdot\left[1+\Gamma\left(\frac{1}{6}\right)\left(\frac{1}{\kappa_m L_0}\right)^{5/3}\sum_{i=0}^{\infty}\frac{1}{i!}\left(\frac{1}{\kappa_m L_0}\right)^{2i}{}_1F_1\left(-i-\frac{5}{6};1;-\frac{\kappa_m^2 r_-^2}{4}\right)\right.$$

$$\left.-\exp\left(\frac{r_-}{L_0}\right){}_1F_1\left(-\frac{1}{3};-\frac{2}{3};-\frac{2r_-}{L_0}\right)\right]. \tag{55}$$

Fig. 11 shows the BER affected by atmospheric turbulence in the downlink when wave beams are received by the large aperture antenna where $a_e = 7.5$ m. It is found that the decrease in the spatial coherence of received waves causes the decrease in the average received power and result in the increase in BER. Note that an influences of spot dancing is negligible because a statistical characteristics of received waves can be considered as a plane wave as mentioned in the introduction of (28)

3.3 Effects of antenna radius of ground station on BER performance

In the system design of the ground station, we may increase an aperture radius of the ground station's antenna in order to satisfy the required Effective Isotropic Radiated Power (EIRP) of the transmitter system or the G/T of the receiver system. In this section, we estimate an effect of increasing an aperture radius of the ground station's antenna on BER affected by atmospheric turbulence.

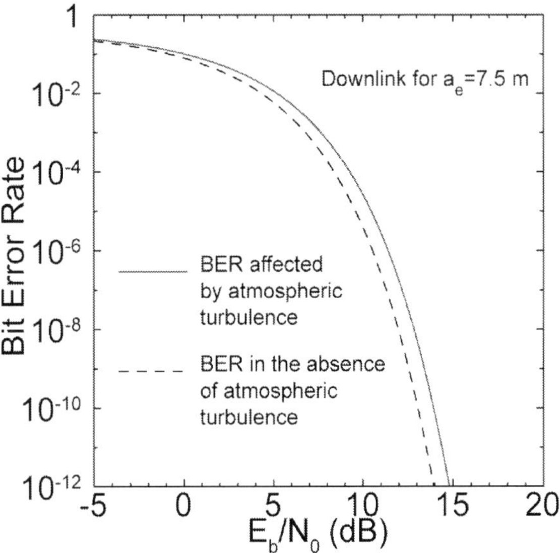

Fig. 11. BER derived from the average received power (PE$_P$) in the downlink in $a_e = 7.5$ m

The EIRP of the transmitter system is defined as the product of a transmitting power and an antenna gain of the transmitting antenna. The transmitting power P_t is obtained by

$$P_t = \frac{1}{Z_0} \cdot \iint_{-\infty}^{\infty} d\mathbf{r} \ |u_{\text{in}}(\mathbf{r}, 0)|^2 = \frac{A}{Z_0}, \tag{56}$$

where $u_{\text{in}}(\mathbf{r}, 0)$ is given by (15). The antenna gain of the transmitting antenna G_t is defined by

$$G_t = \frac{4\pi z_L^2 S}{P_t}, \tag{57}$$

where S denotes the received power density at $(\mathbf{0}, z_L)$:

$$S = \frac{|u(\mathbf{0}, z_L)|^2}{Z_0} = \frac{1}{Z_0} \cdot \frac{2A}{\pi w^2}. \tag{58}$$

Thus, the antenna gain can be expressed by

$$G_t = \frac{4\pi z_L^2 S}{P_t} = \frac{8 z_L^2}{w^2} = \frac{8 z_L^2}{w_0^2(1 + p^2)} \simeq 2(k w_0)^2, \tag{59}$$

where it is assumed that $p^2 = 4 z_L^2 / (k^2 w_0^4) \gg 1$, which is satisfied in this model of analysis. Using (56) and (59), the EIRP of the transmitter system can be described by

$$\text{EIRP} = P_t \cdot G_t = \frac{2A(k w_0)^2}{Z_0}. \tag{60}$$

The G/T of the receiver system can be expressed by the ratio of an antenna gain of the receiving antenna to the system noise temperature. The antenna gain of the receiving antenna G_r can be described by

$$G_r = \frac{4\pi}{\lambda^2} S_e = 4\pi \cdot \left(\frac{k}{2\pi}\right)^2 \cdot \frac{\pi a_e^2}{2} = \frac{(ka_e)^2}{2}, \tag{61}$$

where $S_e = \pi a_e^2/2$ because the aperture efficiency of the receiving antenna, whose field distribution is given by (32), is 0.5. The system noise temperature T_s is obtained by

$$T_s = \frac{N_0}{k_B}, \tag{62}$$

where k_B denotes Boltzmann's Constant. Thus, the G/T of the receiver system can be described by

$$G/T = \frac{G_r}{T_s} = \frac{k_B}{N_0} \cdot \frac{(ka_e)^2}{2}. \tag{63}$$

On the other hand, using (19) and (34), E_b/N_0 in free space is obtained by

$$\frac{E_b}{N_0} = \frac{P_{in}(z_L) \cdot T_b}{N_0} = \frac{T_b}{N_0} \cdot \frac{A}{Z_0} \frac{a_e^2}{w^2 + a_e^2} \simeq \frac{T_b}{k_B T_s} \cdot \frac{A}{Z_0} \frac{a_e^2}{w^2} = \frac{T_b}{k_B T_s} \cdot \frac{A}{Z_0} \cdot a_e^2 \cdot \frac{k^2 w_0^2}{4 z_L^2}, \tag{64}$$

where it is assumed that $a_e/w \ll 1$. Using the EIRP and the G/T obtained by (60) and (63) respectively, E_b/N_0 in free space can be expressed by

$$\begin{aligned}
\frac{E_b}{N_0} &= \frac{T_b}{k_B} \cdot \frac{A}{Z_0} \cdot 2(kw_0)^2 \cdot \frac{1}{(2kz_L)^2} \cdot \frac{(ka_e)^2}{2T_s} = \frac{T_b}{k_B} \cdot P_{in} \cdot G_t \cdot \frac{1}{(2kz_L)^2} \cdot \frac{G_r}{T_s} \\
&= \frac{T_b}{k_B} \cdot \text{EIRP} \cdot \frac{1}{(2kz_L)^2} \cdot G/T.
\end{aligned} \tag{65}$$

Note that $(2kz_L)^2$ is the free space path loss.

3.3.1 Uplink

Using (65), we can describe BER derived from the average received intensity given by (50) in the uplink:

$$\text{PE}_I = \frac{1}{2}\text{erfc}\left(\sqrt{S_I \cdot \frac{T_b}{k_B} \cdot \text{EIRP} \cdot \frac{1}{(2kz_L)^2} \cdot G/T}\right). \tag{66}$$

Fig. 12 shows the BER as a function of kw_0 under the condition that G/T and EIRP keep constant, where the transmitting power A/Z_0 changes in inverse proportion to the square of kw_0 in (60). It is found that the BER affected by atmospheric turbulence increases as kw_0 becomes large, whereas the BER in the absence of atmospheric turbulence plotted by the dashed line does not change. Fig. 13 shows the BER for various beam radius at the transmitting antenna w_0 as a function of E_b/N_0 obtained by (65). It is shown that BER increases as w_0 becomes larger as well as Fig. 12.

The reason for the increase in BER is as follows.

We have shown that spot dancing of wave beams due to atmospheric turbulence causes the increase in BER for the uplink in Sec. 3.2.1. From each profile of the intensity in the absence of atmospheric turbulence plotted by the dashed line in Figs. 14 to 17, it is found that the

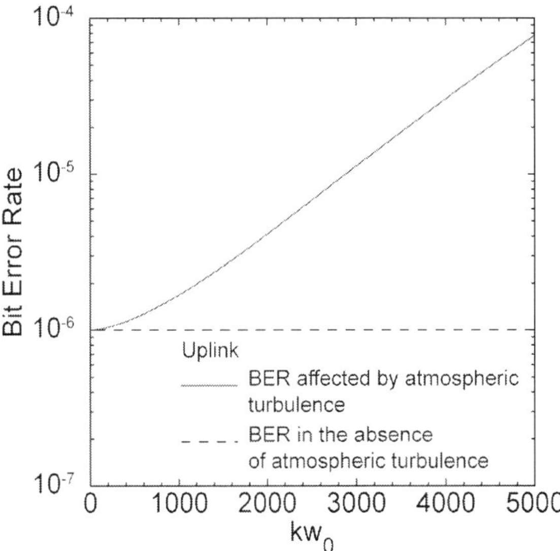

Fig. 12. BER derived from the average received intensity in the uplink as a function of kw_0 when the EIRP of the transmitter system keeps constant.

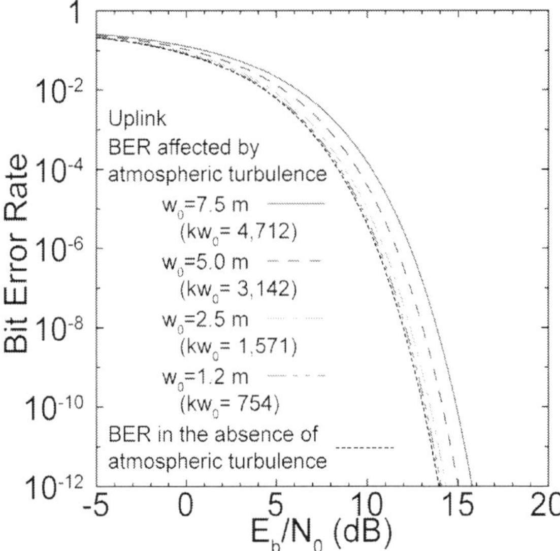

Fig. 13. BER derived from the average received intensity in the uplink for various beam radius at the transmitting antenna w_0 as a function of E_b/N_0.

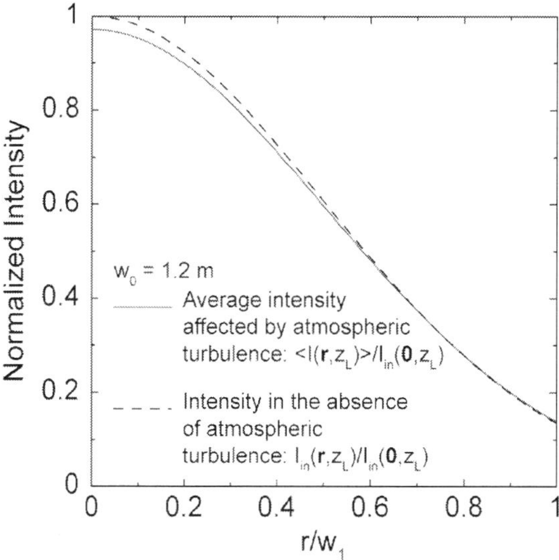

Fig. 14. Average intensity for the uplink in the beam radius at the transmitting antenna $w_0 = 1.2$ m normalized by the intensity on a beam axis in free space as a function of the distance from the center of the receiving antenna scaled by w_1, which denotes the beam radius at the plain of the receiving antenna for $w_0 = 1.2$ m.

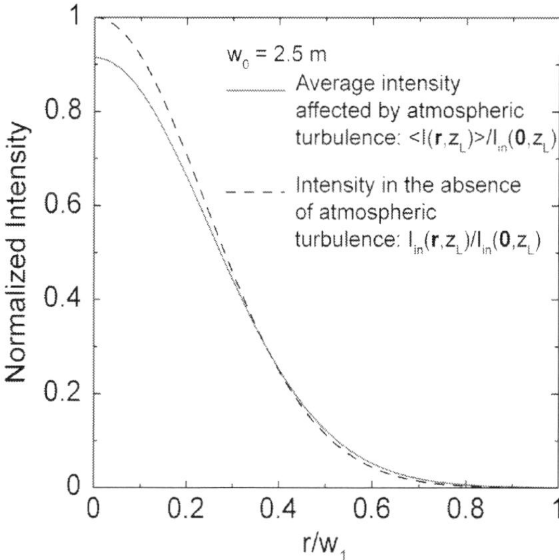

Fig. 15. Same as Fig 14 except for $w_0 = 2.5$ m.

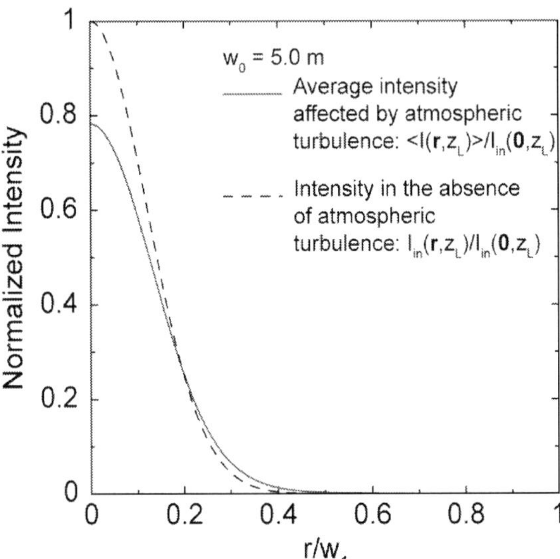

Fig. 16. Same as Fig 14 except for $w_0 = 5.0$ m.

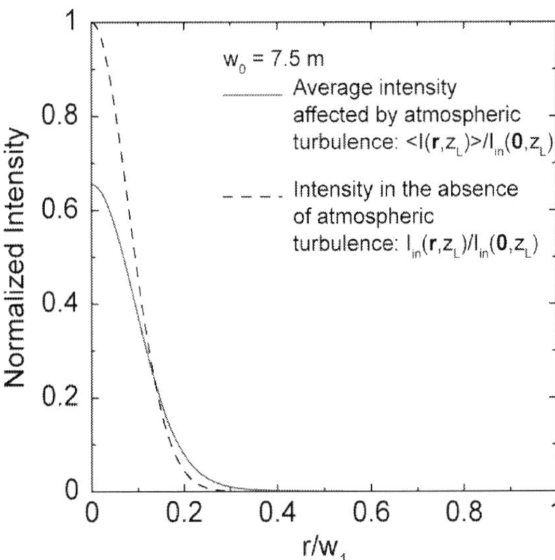

Fig. 17. Same as Fig 14 except for $w_0 = 7.5$ m.

beam spot size at the plain of the receiving antenna becomes smaller as w_0 increases. The displacement of the arrived beam axis due to spot dancing makes the received intensity decrease considerably faster as the beam spot size becomes smaller. Therefore, the average intensity affected by atmospheric turbulence decreases at the center of the receiving antenna and the profile is spread as w_0 increases as shown in Figs. 14 to 17. This is why BER in the uplink increases as an aperture radius of the ground station's antenna becomes larger.

From these results, we find that the increase in the transmitting power is better than the increase in the aperture radius of the ground station's antenna in order to satisfy the required EIRP from the point of view of the decrease in an influence of atmospheric turbulence on BER in the uplink.

3.3.2 Downlink

For the downlink, we can obtain BER derived from the average received power given by (53):

$$\mathrm{PE_P} = \frac{1}{2}\mathrm{erfc}\left(\sqrt{S_P \cdot \frac{T_b}{k_B} \cdot \mathrm{EIRP} \cdot \frac{1}{(2kz_L)^2} \cdot G/T}\right). \tag{67}$$

Fig. 18 shows the BER as a function of ka_e under the condition that EIRP and G/T keep constant, where the noise power density N_0 changes in inverse proportion to the square of ka_e in (63). It is found that the BER affected by atmospheric turbulence increases as ka_e becomes larger. Fig. 19 shows the BER for various aperture radius of the receiving antenna as a function of E_b/N_0 obtained by (65). It is shown that BER increases as a_e becomes larger as well as Fig. 18. From results of the DOC in Fig. 9, it is found that the spatial coherence radius becomes smaller relative to a radius of the receiving antenna and then the spatial coherence

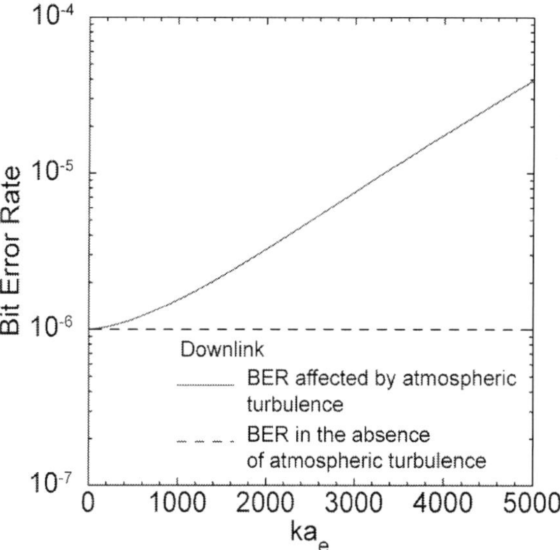

Fig. 18. BER derived from the average received power in the downlink as a function of ka_e when the G/T of the receiver system keeps constant.

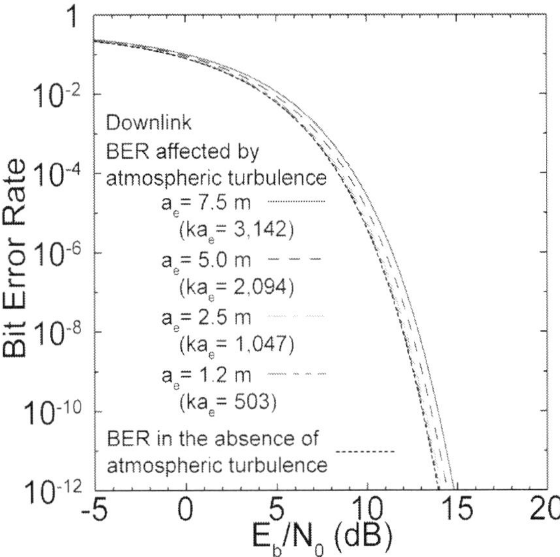

Fig. 19. BER derived from the average received power in the downlink for various aperture radius of the receiving antenna a_e as a function of E_b/N_0.

of received waves decreases as the radius of the antenna increases. The effect of the spatial coherence of received waves causes the decrease in the average received power and results in the degradation of BER performance.

From these results, it is found that the decrease in the system noise temperature by the improvement of a receiver's noise is better than the increase in an aperture radius of the ground station's antenna in order to decrease an influence of atmospheric turbulence on BER for the downlink in the design to satisfy the required G/T.

4. Conclusion

We analyzed BER derived from the average received power, which is deduced by the second moment of a Gaussian wave beam, for the GEO satellite communications in Ka-band at low elevation angles affected by atmospheric turbulence. We find the followings:

1. For the uplink, the decrease in the average received intensity caused by spot dancing of wave beams degrades the BER performance. However, the spatial coherence of received wave beams decreases little and there are little influences of this spatial coherence on BER.

2. For the downlink, the decrease in the spatial coherence of received wave beams degrades the BER performance. However, spot dancing of wave beams influences little on BER.

3. In the design of the ground station, the increase in a transmitting power for the uplink or the decrease in the noise temperature of the receiver system for the downlink is better than the increase in an aperture radius of the ground station's antenna in order to satisfy the required EIRP of the transmitter system or G/T of the receiver system from the point of view of the decrease in an influence of atmospheric turbulence on BER performance.

In this chapter, we do not consider effects of the higher moment of a Gaussian wave beams on BER. At the next stage, we will analyze effects of the fourth moment of received wave beams on BER for the GEO satellite communications. Furthermore, we have to consider the probability density function (PDF) about the bit error of satellite communications affected by atmospheric turbulence in order to make a more actual analysis. An introduction of the PDF is a future problem.

5. References

Andrews, L. C. & Phillips, R. L. (2005). *Laser Beam Propagation through Random Media*, 2nd edn, SPIE Press.

Fante, R. L. (1975). Electromagnetic beam propagation in turbulent media, *Proceedings of the IEEE* 63(12): 1669–1692.

Fante, R. L. (1980). Electromagnetic beam propagation in turbulent media: An update, *Proceedings of the IEEE* 68(11): 1424–1443.

Hanada, T., Fujisaki, K. & Tateiba, M. (2008a). Theoretical analysis of bit error rate of satellite communication in Ka-band under spot dancing and decrease in spatial coherence caused by atmospheric turbulence, *Progress In Electromagnetics Research C* 3: 225–245.

Hanada, T., Fujisaki, K. & Tateiba, M. (2008b). Theoretical analysis of bit error rate of satellite communications in Ka-band through atmospheric turbulence, *Proceedings of the 7th Asia-Pacific Engineering Research Forum on Microwaves and Electromagnetic Theory*, Fukuoka Institute of Technology, Fukuoka, Japan, pp. 7–13.

Hanada, T., Fujisaki, K. & Tateiba, M. (2009a). Average bit error rate for satellite downlink communications in Ka-band under atmospheric turbulence given by Gaussian model, *Proceedings of 2009 Asia-Pacific Microwave Conference (APMC 2009)*, Singapore.

Hanada, T., Fujisaki, K. & Tateiba, M. (2009b). Average bit error rate for satellite uplink communications in Ka-band under atmospheric turbulence given by Gaussian model, *Proceedings of the 15th Asia-Pacific Conference on Communications (APCC 2009)*, Shanghai, China, pp. 438–441.

Hanada, T., Fujisaki, K. & Tateiba, M. (2009c). Theoretical analysis of bit error rate for downlink satellite communications in Ka-band through atmospheric turbulence using Gaussian model, *Proceedings of 2009 Korea-Japan Joint Conference on AP/EMC/EMT*, Incheon, Korea, pp. 35–38.

Hanada, T., Fujisaki, K. & Tateiba, M. (2009d). Theoretical analysis of bit error rate for satellite communications in Ka-band under atmospheric turbulence given by Kolmogorov model, *Journal of Electromagnetic Waves and Applications* 23(11–12): 1515–1524.

Ippolito, L. J. (2008). *Satellite Communications Systems Engineering: Atmospheric Effects, Satellite Link Design and System Performance*, John Wiley and Sons, Ltd.

Ishimaru, A. (1997). *Wave Propagation and Scattering in Random Media*, IEEE Press and Oxford University Press.

Karasawa, Y., Yamada, M. & Allnutt, J. E. (1988). A new prediction method for tropospheric scintillation on earth-space paths, *IEEE Transactions on Antennas and Propagation* 36(11): 1608–1614.

Karasawa, Y., Yasukawa, K. & Yamada, M. (1988). Tropospheric scintillation in the 14/11-GHz bands on earth-space paths with low elevation angles, *IEEE Transactions on Antennas and Propagation* 36(4): 563–569.

Martini, E., Freni, A., Facheris, L. & Cuccoli, F. (2006). Impact of tropospheric scintillation in the Ku/K bands on the communications between two LEO satellites in a radio occultation geometry, *IEEE Transactions on Geoscience and Remote Sensing* 44(8): 2063–2071.

Marzano, F. S., Riva, C., Banich, A. & Clivio, F. (1999). Assessment of model-based scintillation variance prediction on long-term basis using Italsat satellite measurements, *International Journal of Satellite Communications* 17: 17–36.

Matricciani, E., Mauri, M. & Riva, C. (1997). Scintillation and simultaneous rain attenuation at 12.5 GHz to satellite Olympus, *Radio Science* 32(5): 1861–1866.

Matricciani, E. & Riva, C. (2008). 18.7 GHz tropospheric scintillation and simultaneous rain attenuation measured at Spino d'Adda and Darmstadt with Italsat, *Radio Science* 43.

Mayer, C. E., Jaeger, B. E., Crane, R. K. & Wang, X. (1997). Ka-band scintillations: Measurements and model predictions, *Proceedings of the IEEE* 85(6): 936–945.

Otung, I. E. (1996). Prediction of tropospheric amplitude scintillation on a satellite link, *IEEE Transactions on Antennas and Propagation* 44(12): 1600–1608.

Otung, I. E. & Savvaris, A. (2003). Observed frequency scaling of amplitude scintillation at 20, 40, and 50 GHz, *IEEE Transactions on Antennas and Propagation* 51(12): 3259–3267.

Peeters, G., Marzano, F. S., d'Auria, G., Riva, C. & Vanhoenacker-Janvier, D. (1997). Evaluation of statistical models for clear-air scintillation prediction using OLYMPUS satellite measurements, *International Journal of Satellite Communications* 15: 73–88.

Rytov, S. M., Kravtsov, Y. A. & Tatarskii, V. I. (1989). *Principle of Statistical Radiophysics 4 Wave Propagation through Random Media*, Springer-Verlag.

Strohbehn, J. W. (ed.) (1977). *Laser Beam Propagation in the Atmosphere*, in Topics in Applied Physics, Springer-Verlag, Berlin and New York.

Tatarskii, V. I. (1961). *Wave Propagation in a Turbulent Medium*, McGraw-Hill, New York.

Tatarskii, V. I. (1971). *The Effects of the Turbulent Atmosphere on Wave Propagation*, Israel Program for Scientific Translations, Jerusalem.

Tatarskii, V. I., Ishimaru, A. & Zavorotny, V. U. (eds) (1993). *Wave Propagation in Random Media (Scintillation)*, The Society of Photo-Optical Instrumentation Engineers and IOP Publishing Ltd.

Tateiba, M. (1974). Moment equation of a wave propagating through random media, *Memoirs of the Faculty of Engineering, Kyushu University* 33(4): 129–137.

Tateiba, M. (1975). Mechanism of spot dancing, *IEEE Transactions on Antennas and Propagation* AP-23(4): 493–496.

Tateiba, M. (1982). Multiple scattering analysis of optical wave propagation through inhomogeneous random media, *Radio Science* 17(1): 205–210.

Tateiba, M. (1985). Some useful expression for spatial coherence functions propagated through random media, *Radio Science* 20(5): 1019–1024.

Uscinski, B. J. (1977). *The Elements of Wave Propagation in Random Media*, McGraw-Hill, Inc.

Vasseur, H. (1999). Prediction of tropospheric scintillation on satellite links from radiosonde data, *IEEE Transactions on Antennas and Propagation* AP-47(2): 293–301.

Wang, T. & Strohbehn, J. W. (1974). Log-normal paradox in atmospheric scintillations, *Journal of the Optical Society of America* 64(5): 583–591.

Wheelon, A. D. (2003). *Electromagnetic Scintillation II. Weak Scattering*, Cambridge University Press.

Part 3

Real Time Applications over Satellite

3

Improving Quality-of-Service of Real-Time Applications over Bandwidth Limited Satellite Communication Networks via Compression

LingSun Tan, SeiPing Lau and ChongEng Tan
Universiti Malaysia Sarawak,
Malaysia

1. Introduction

VSAT (Very Small Aperture Terminal) satellite network is one of the widely deployed communication networks for rural and remote communications in today's telecommunication world. VSAT satellite networks are growing steadily throughout many industries and market segments in many countries. With new applications and shifts in target markets, VSAT based solutions are being adopted at increasingly higher rates since year 2002 (MindBranch, 2011). Up to December 2008, VSAT market statistics show that the total number of Enterprise VSAT terminals being ordered is 2,276,348, the total number of VSATs being shipped is 2,220,280 and the total number of VSAT sites in service is 1,271,900 throughout the world (Comsys, 2008). VSAT satellite network offers value-added satellite-based services capable of supporting the Internet, data, video, LAN, voice and fax communications. VSATs are a single, flexible communication platform which can be installed quickly and cost efficiently to provide telecommunication solutions for consumers, governments and corporations, thus, they are becoming increasingly important.

VSAT satellite network plays an important role in bridging the digital divide and it is the one of the easiest deployment technology and cost effective way to interconnect two networks especially in rural areas, when other wired technologies are practically impossible and unsuitable due to geographical distance or accessibility. In this chapter, a fundamental overview of satellite communication network, with the highlighting of its main characteristics, constraints and proposal on compression technique which can be applied to boost up the Quality of Service (QoS) of the satellite communication services, are provided. VSAT satellite network provides communications support for a wide range of applications, which include point-of-sales transaction, financial management, telemetry & data collection, private-line voice services, virtual private networks, distance education, high speed internet access and more (TM, 2011).

VSAT satellite technology has many advantages. It can be deployed anywhere around the world and it offers borderless communication within the coverage area. Besides, it is cost effective and can be setup in a matter of minutes. VSAT network configuration such as bandwidth, interfaces and data rates can be updated remotely from the central network management system, hence, it provides high flexibility and efficiency. However, like other technologies, VSAT satellite network has its downsides. The limitations of VSAT technology include the extremely high start-up cost needed for building and launching satellites in the geosynchronous orbit, high round-trip latency of about 500 ms as it utilise the satellites in geosynchronous orbit, and rain attenuation might affect the performance of VSAT communications under rainy conditions (TopBits.com, 2011). Moreover, it provides low and limited network bandwidth resulting in network congestion, reduced Quality-of-Service (QoS) of real-time interactive multimedia applications and also late packet delivery issues. These issues have created some negative impacts on the QoS of communication networks and also user experiences.

Apart from the need for efficient mechanisms for storage and transfer of enormous volume of data, these also lead to insatiable demands for ever-greater bandwidth in VSAT satellite network. In order to strike a balance between the cost and offered satellite bandwidth, some enhancements have to be implemented to reduce the bandwidth requirement of real-time applications that demanding high bandwidth and fully optimize the use of the low speed satellite link. Several techniques have been introduced to further improve the network bandwidth utilization and reduce network traffic especially for wireless satellite networks (Tan et al., 2010). One of such techniques is via compression, which is a technique used to overcome the network packet overhead by eliminating redundancies in packet delivery. By reducing the packet size, more packets can be transmitted over the same communication link at one time and hence increase the efficiency of bandwidth utilization. In this chapter, the concept of data compression is examined in order to know in depth how data compression can actually play a role in improving user experience. After that, the basic concept of packet compression, which consists of header compression and payload compression is also discussed.

Currently, there are many compression schemes, systems and frameworks have been proposed and designed in order to perform efficient data compression for better utilization of the communication channel. However, most of them have their own advantages and limitations, which may not suit for VSAT satellite network environment. For example, the Adaptive Compression Environment (ACE) system which has been proposed might impose additional delays over VSAT satellite network due to computation overhead and large compression time cost of the algorithm used. Besides, the Adaptive Online Compression (AdOC) algorithm which is proposed in the related work might cause the satellite link to be more congested due to the increased network load caused by the algorithm. In addition, some of the proposed compression schemes are designed for a specific aspect, which might create additional issues working under VSAT satellite network. Thus, in this chapter, the performance of several well-known compression schemes are reviewed and evaluated under the context of bandwidth limited VSAT satellite network, in order to highlight important criterions for improving performance over low bandwidth VSAT satellite network. Finally, the proposed enhanced compression scheme will be presented and the performance of the compression scheme will be examined and evaluated through extensive network simulations.

2. Introduction to VSAT communication

VSAT satellite network has become an essential part of our daily lives in recent years. It is used widely in telephony communication, broadband and internet services, and military communication. VSAT is a small satellite dish that is capable of both receiving and sending satellite signals (TM, 2011). It can be used for two-way communications via satellite.

Generally, satellite is a specialized wireless receiver or transmitter that is launched by a rocket and placed in orbit around the earth (DotNetNuke Corporation, 2010). Thus, it is capable of providing coverage over large geographical areas and establishing communication links between various points on earth.

2.1 Basic satellite elements

Satellite communication system is comprised of two main components, namely space segment and ground segment, as illustrated in Figure 1 below. A basic satellite communication system consists of a space segment serving a specific ground segment (Richharia, 1999). The satellite itself is also known as the space segment while the earth stations will serve as the ground segment. The satellite is controlled and its performance is monitored by the Telemetry Tracking and Command (TT&C) station.

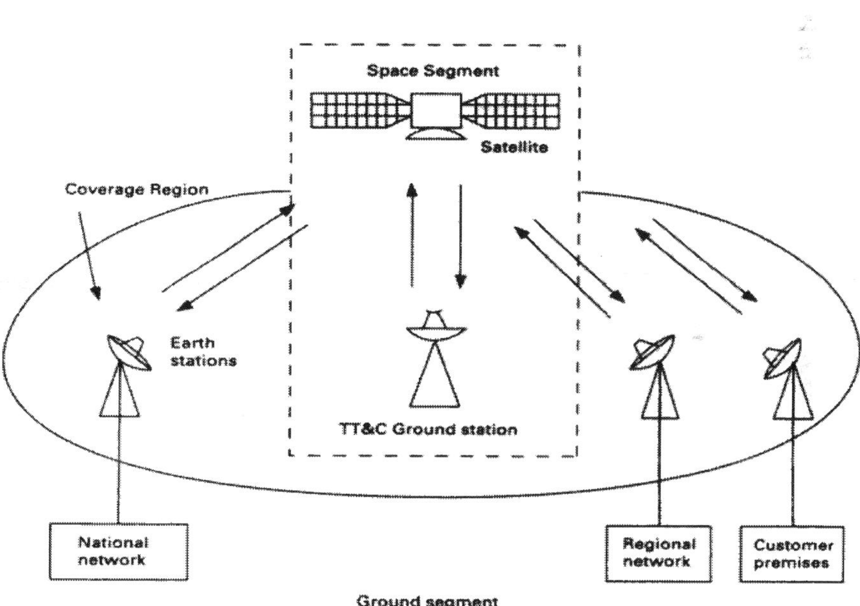

Fig. 1. The main elements of a satellite communication network (Richharia, 1999).

Communication can be established easily between all earth stations located within the coverage region through the satellite. The primary role of a satellite is to relay electronic signals. When signals from the earth stations are received by the satellite, the signals are processed, translated into another radio frequency and retransmitted down towards another desired earth stations after further amplification. Satellite relay can be two way, as in the case of a long distance phone call, and point to multipoint, as in the case with television broadcasts.

2.2 Satellite roles and applications

The most important role of satellite communication network is to provide connectivity to the user terminals and to internetwork with terrestrial networks so that the applications and services provided by terrestrial network such as telephony, television, broadband access and Internet connections can be extended to places where cable and terrestrial radio cannot economically be installed and maintained. Satellite network provides direct connections among user terminals, connections for terminals to access terrestrial networks and connections between terrestrial networks (Mitra, 2005).

Since satellite is capable of providing coverage over a much wider area such as oceans, intercontinental flight corridors and large expanses of land mass, it is used in providing voice and data communications to aircraft, ships, land vehicles and handsets. Besides, satellite allows passengers on an aircraft to connect directly to a land based telecommunication network. Apart from that, it is also used for remote sensing, earth observation, meteorological applications such as weather survey, military communication and global positioning services (GPS).

2.3 Limitations of satellite communication

Three main characteristics and constraints of satellite network are high latency, poor bandwidth and noise (Hart, 1997). High latency is one of the main limitations of satellite network and it is caused by the long propagation path due to the high altitude of satellite orbits. In satellite network, the time required to navigate through a satellite link is longer compared to terrestrial network. Hence, this leads to higher transmission delay.

For geostationary (GEO) satellite communication system, the time required to traverse these distances, namely, earth station to satellite, then satellite to another earth station, is around 250ms (Sun, 2005). Round-trip delay will be 500ms. These propagation times are much greater than those encountered in conventional terrestrial systems. The high latency constraint of satellite link might not affect bulk data transfer and broadcast-type applications, but it will affect those highly interactive real-time applications.

Due to radio spectrum limitations, satellite transmission has a fixed amount of bandwidth (Hart, 1997). Problems like network congestion and packet loss might occur when those real-time interactive applications that consume high bandwidth are running over satellite link. Furthermore, strength of radio signal is in proportion to the square of distance traveled (Hart, 1997). Thus, signals traverse through satellite link might get very weak due to long distance between earth stations and satellite.

3. Data compression

Data compression plays an important role in improving the performance of low bandwidth VSAT satellite network. Among other satellite performance enhancement techniques, data compression is the most suitable and economical way to further improve the user experience of VSAT satellite network. This is because data compression technique is much simpler and can be implemented easily. Currently, a lot of the networking corporations are providing solutions for improving Internet services over satellite network by using high cost network equipments. These products are very costly and require complicated hardware configuration, while data compression is freely available and no complicated hardware configuration is required. Thus, data compression is adopted in the proposed scheme. Lately, data compression has become a common requirement for most application software as well as an important and active research area in computer science study. None of the ever-growing Internet, digital television, mobile communication or increasing video communication techniques would have been the practical developments without applying compression techniques.

In general, data compression is a process of representing information in a more compact form by eliminating redundancies in the original data representation (Pu, 2006). Due to the presence of redundancies in the original representation, data such as text, image, sound or any combination of all these types such as video is not in the shortest form, thus rendering its compression a possibility. Data compression is adopted in a variety of application areas such as mobile computing, image archival, video-conferencing, computer networks, digital and satellite television, multimedia evolution, imaging and signal processing. It can be divided into two major categories, namely lossless and lossy compression.

3.1 Lossless compression

In lossless compression, the exact original data can be reconstructed from the compressed data without any loss of information (Pu, 2006). Each compress-decompress cycle will generate exactly similar data, hence, lossless compression is known as *reversible* compression. Lossless compression techniques are used when storing medical images, text and images preserved for legal reason, some computer executable files, database records, spreadsheets or word processing files, where the lost of even a single bit could be catastrophic.

Example of lossless data compression is shown in Figure 2, where the exact input string FFMMMF is reconstructed after the execution of the compression algorithm followed by the decompression algorithm.

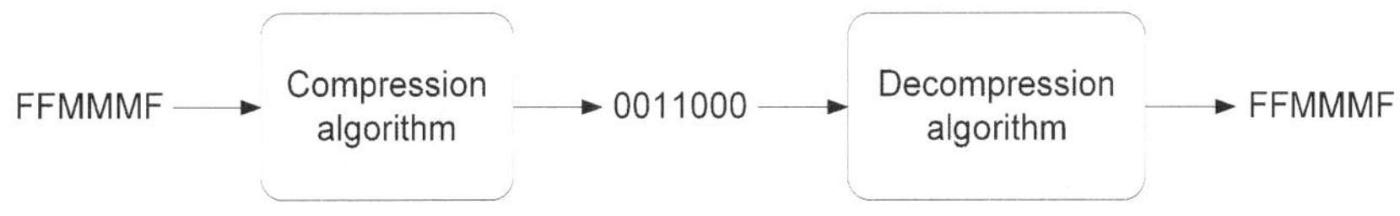

Fig. 2. Example of lossless data compression (Pu, 2006).

3.2 Lossy compression

Lossy compression concedes a certain loss of accuracy in exchanges for greatly improved or more effective compression ratio. Owning to that, it does not allow the exact original data to be reconstructed from the compressed data (Pu, 2006). It usually suffers from information loss as compressing and decompressing the file repeatedly will cause loss of quality gradually. Thus, lossy compression is also known as *irreversible* compression. Lossy compression is used frequently in streaming media and telephony applications as it is proven to be effective over graphic images and digitized voice.

Lossy compression is not suitable for compressing text file formats due to loss of accuracy. However, it produces a much smaller compressed file than any known lossless compression method as it accepts some loss of data in order to achieve higher compression ratio.

Figure 3 shows an example of lossy data compression, where a long decimal number becomes a shorter approximation number after the compression-decompression process.

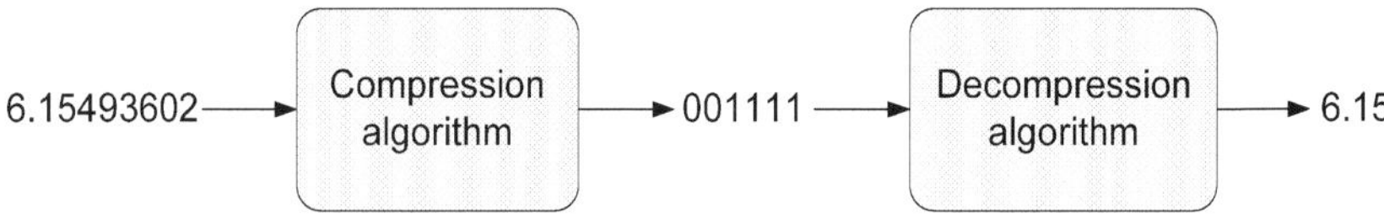

Fig. 3. Example of lossy data compression (Pu, 2006).

4. Reviews of existing works

There are numerous compression schemes, systems and frameworks have been proposed and designed in order to improve the performance of communication network. The networking community has approached the problem of compressing network data streams as packets come by (online scenario), while the database community has focused more on applying such techniques to save storage space (offline scenario) (Chen et al., 2008). However, most of them have their own characteristics, which may not be suitable for satellite network environment. This section briefly discusses on packet compression, header compression and payload compression. Several well known compression schemes are also evaluated for their implementation under satellite network environment. These compression schemes can be further divided into three categories, namely packet compression schemes, header compression schemes and payload compression schemes.

4.1 Packet compression

Data compression in packet network is known as packet compression. Normally in computer network, network data will be divided into smaller chunks before transmission and transmitted as packets over the communication channel. Packet compression allows much smaller amounts of packet drops, more simultaneous sessions, and a smooth and fast behavior of applications (Matias & Refua, 2005). Since network packet consists of two parts, namely header and payload, as shown in Figure 4, therefore packet compression can be achieved by either header or payload compression, or the combination of both. Figure 5 depicts a basic packet compression.

Network packet

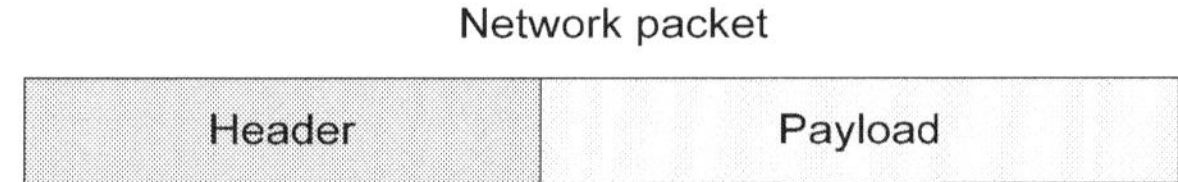

Fig. 4. Structure of a network packet.

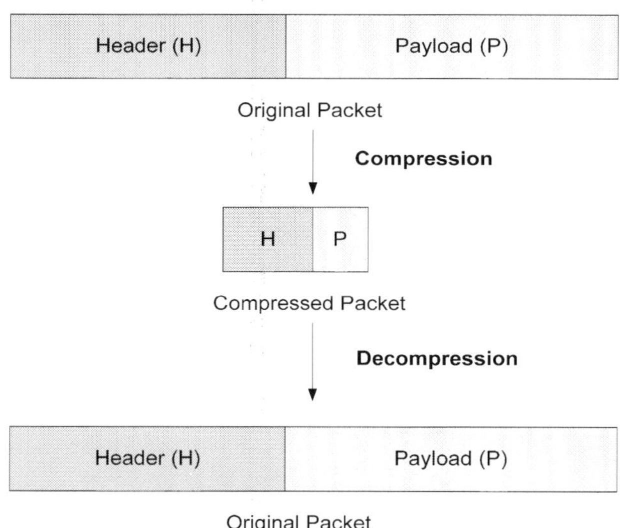

Fig. 5. Basic packet compression.

4.1.1 Packet compression schemes

Packet compression is proposed by some of the related works, as discussed in the following sections.

4.1.1.1 IPzip

A comprehensive suite of algorithms known as IPzip is presented for network packet headers and payloads compression. IPzip is designed to exploit the hidden intra-packet correlation and inter-packet correlation properties of the data streams (Chen et al., 2008). After that, it produces an efficient compression plan, where the data streams both within and across packets are reorganized to improve the compression ratio. The compression plan is built in an offline phase as reordering of packets and fields is resource intensive.

IPzip learns the correlation pattern over a training set, after that generates a compression plan and then compresses the original data set according to the plan. However, the performance of the current compression plan may decrease and new compression plan is needed due to the changes in the intrinsic network traffic pattern. Thus, the effectiveness of the compression plan will be monitored over time. Block compression is introduced, as IPzip aggregates similar packets into a block based on flow information before undergoing compression in order to increase compression ratio.

Unfortunately, IPzip may not suit for real-time processing as it needs to carry out offline training to produce the efficient compression plan. Besides, IPzip may not be able to react if the intrinsic network traffic pattern changes frequently, since the learning process to generate a new compression plan takes time and efforts. Moreover, IPzip will simple cause network congestion if the compression processing speed is slower than the relay processing speed as it compresses all blocks. In conclusion, IPzip is not suitable for satellite network environment.

4.1.1.2 Adaptive packet compression scheme for advanced relay node

This research work presented an adaptive lossless packet compression scheme especially for advanced relay node in the network. This scheme is proposed to mitigate network traffic congestion issue and it is based on the assumption that both the conventional packet delay and additional advanced functions can be performed by the intermediate nodes inside the network using computational or storage resources at the nodes (Shimamura et al., 2009).

This scheme compresses the incoming packets adaptively and selectively according to some important metrics. Packet by packet compression is performed to evaluate the potential of adaptive packet compression inside network.

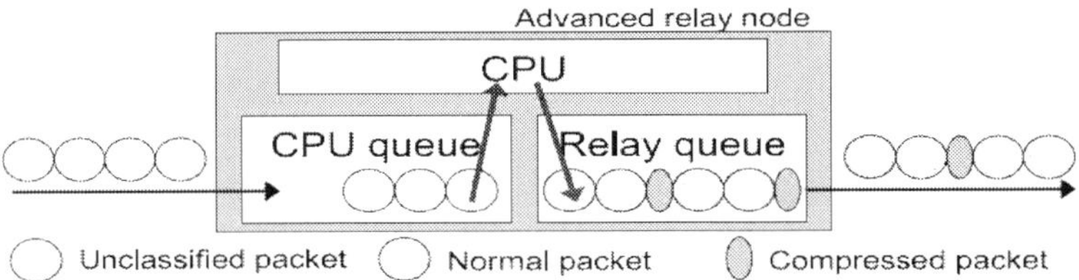

Fig. 6. Adaptive compression model in advanced relay node (Shimamura et al., 2009).

Figure 6 shows the model of adaptive compression scheme proposed for advanced relay node. Notice that an advanced relay node has two logical queues, which are CPU queue and relay queue. When the advanced relay node receives a packet, this compression scheme determines whether packet compression is effective according to the waiting time in the relay queue, compression processing time, packet size, output link bandwidth and compression ratio. If packet compression is beneficial, then compression will be performed on the packet by the advanced relay node before the packet moves to the relay queue.

Simulation results show that this scheme succeeds in reducing the packet delay and packet discard rate. However, the compression ratio achieved via per packet compression is very much lower compared to the one with block compression. Therefore, this scheme will not help much in bandwidth saving when working under low bandwidth satellite link. In addition, massive computation which consume a lot of time needs to be done by the advanced relay node each time it receives a packet. Under a heavy traffic condition which is usually experienced in a low bandwidth satellite link, more and more calculation need to be carried out, thus, more and more delays being created, and finally creates a bad impact on user experience.

4.2 Header compression

The applicability of Internet technology over low speed and high delay links is threatened and reduced by large and repetitive packet headers. Some delay sensitive applications, such as remote login and real-time interactive multimedia applications, need to use small packets (Naidu & Tapadiya, 2009). However, the overhead of large packet headers on small packets can be prohibitive. A natural way to alleviate the problem is to compress packet header as packet header information shows significant redundancy between consecutive packets. Header compression makes more efficient use of link bandwidth in a packet switched network by leveraging header field redundancies in packets belonging to the same packet stream (Taylor et al., 2005).

Most of the header fields such as source and destination address remain constant throughout the duration of a flow, while other fields such as sequence numbers change predictably. Thus, the header size can be significantly reduced for most packets by sending static fields information only initially and utilizing dependencies and predictability for other fields. The reference copies of full headers must be stored at the context of compression and decompression sides in order to communicate and reconstruct the original packet headers reliably.

Initially, a few packets are sent uncompressed and they are used to establish the shared state called context on both sides of the link. The context comprises information about static fields, dynamic fields and their change pattern in protocol headers. The compressor will use this information to compress the packet as efficiently as possible and then the decompressor will decompress the packet to its original state. To correctly decompress the compressed packet header, synchronization between compressor and decompressor is mandatory.

4.2.1 Header compression schemes

In order to overcome packet header overhead, two popular header compression schemes have been developed. They are Van Jacobson header Compression (VJHC) scheme and RObust Header Compression (ROHC) scheme.

4.2.1.1 Van Jacobson Header Compression (VJHC)

Van Jacobson Header Compression scheme was introduced by Jacobson in year 1990. This scheme is used to improve the interactive terminal response of low speed serial modem links and it is specially developed for Transmission Control Protocol/Internet Protocol (TCP/IP) (Jacobson, 1990). VJHC is commonly used to compress the header of IPv4/TCP packets.

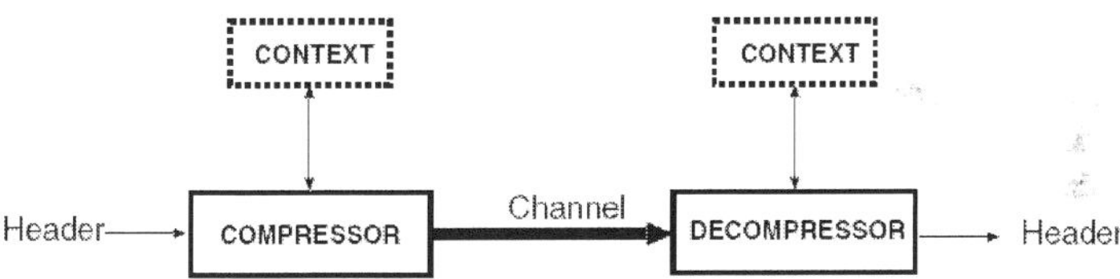

Fig. 7. The flow context concept in header compression (Suryavanshi et al., 2004).

As shown in Figure 7, the concept of flow context is used during the process of header compression. For each packet flow, the context is built on both the compressor and decompressor side and a unique context identity (CID) is assigned. The flow context information is made up of a collection of field values and change patterns of field values in the packet header.

To establish the context, first few packets of a newly identified flow are sent to the decompressor without compression. This is because the packet header information needed to be stored for future reference. Once the context is formed on both sides, the compressor starts compressing the packets and now only the encoded difference to the preceeding header is transmitted. When the compressed packets reach the decompressor side, the changes contained in the newly received compressed header is applied to the saved header in the context to obtain the uncompressed headers.

VJHC scheme stated that TCP/IP header fields can be grouped into several categories, which are constant, inferred and dynamic (Jacobson, 1990). Constant fields are those field values that remained unchanged between consecutive packets, hence, can be eliminated. Inferred fields are those fields that can be recalculated at the receiving end. For example, 'total length' and 'header checksum' field. The transmission efficiency can be improved significantly by suppressing inferred fields at the compressor and restoring them at the decompressor. The third group is dynamic fields which do no change frequently at the same time or change slightly, thus it can be omitted in most cases.

VJHC is proven to be effective towards header compression, as it can reduce TCP/IPv4 header from 40B to 4B, which is 10% of its original size (Tye & Fairhurst, 2003). However, the main disadvantage of VJHC scheme is it may lead to error propagation throughout the transmission when a compressed packet is lost on the link. This is due to the inconsistent context which will cause a series of packets to be discarded at the receiver end. Thus, VJHC scheme is not applicable under satellite link with high bit error rate as this will lead to higher packet drop which will cause the satellite link performance to become even worse.

4.2.1.2 RObust Header Compression (ROHC)

Besides VJHC, RObust Header Compression (ROHC) scheme is another well known header compression scheme. It is developed by ROHC working group of the IEFT (Tye & Fairhurst, 2003). ROHC is used for compressing IP packet headers and it is particularly suitable for wireless network. ROHC scheme allows bandwidth savings up to 60% in VOIP and multimedia communication applications (JCP-Consult, 2008). In this scheme, compression and decompression are treated as a series of states.

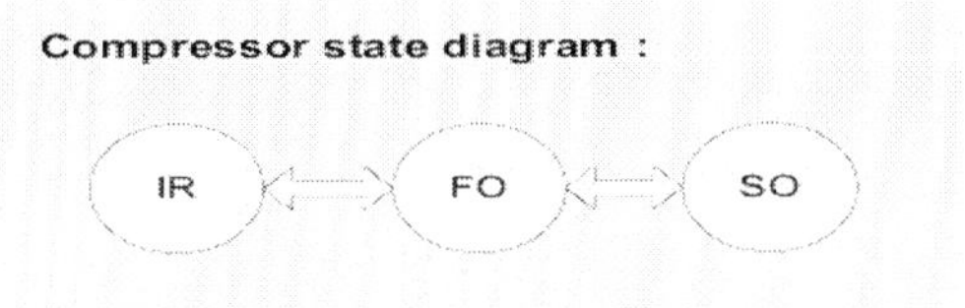

Fig. 8. Compressor state diagram (Effnet, 2004).

As shown in Figure 8, ROHC compressor operates in 3 states, which are Initialization and Refresh (IR), First Order (FO) and Second Order (SO) (Effnet, 2004). The concept of flow context is also adopted in this scheme. The states describe the increasing level of confidence about the correctness of the context at the decompressor side. This confidence is reflected in the increasing compression of packet headers. Initially, the compressor will start with the lowest state and gradually moving to higher state. When there is any error occurred, which will be indicated in the feedback packets, the compressor will move to a lower state to resend packets to fix the error.

Similar to the compressor, ROHC decompressor also operates in 3 states, namely No Context, Static Context and Full Context as illustrated in Figure 9 below (Effnet, 2004). In the beginning of the packet flow, the decompressor will start in the first state, No Context as it has no context information available yet. Once the context information is created at the decompressor site, the decompressor will move to higher state, Full Context state. In the case of error condition, the decompressor will move to lower state to fix the error.

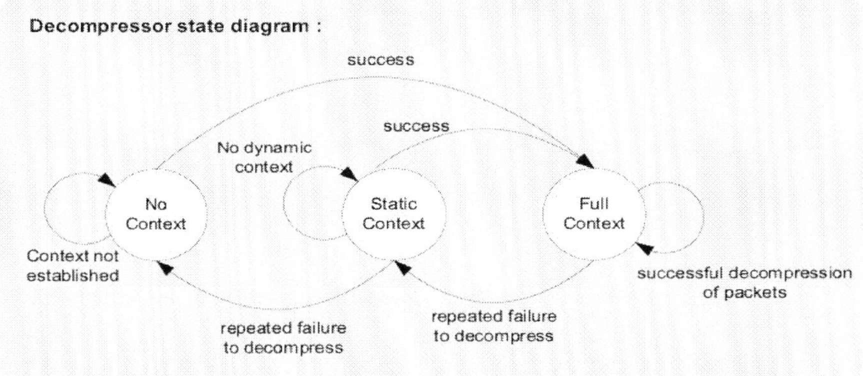

Fig. 9. Decompressor state diagram (Effnet, 2004).

The major advantages of ROHC over VJHC are improved efficiency and high robustness. ROHC works well over links with high bit error rates and long round trip times such as cellular and satellite network. Moreover, its framework is extensible and it is designed to discover dependencies in packets from the same packet flow. However, ROHC scheme is very complicated to be implemented as it absorbed all the existing compression techniques. In addition, in ROHC scheme, the decompressor needs to generate feedback packet and send it back to the compressor to acknowledge successful decompression. Besides that, the context updating information is also sent periodically to ensure context synchronization. This will easily lead to network congestion when working under a low bandwidth satellite link with heavy traffic flows as ROHC scheme increases the network load by generating feedback and context information packets from time to time.

4.3 Payload compression

Packet payload is used to store user information and bulk compression method is usually used for compressing packet payload. Bulk compression treats information in the packets as a block of information and compresses it using a compression algorithm (Tye & Fairhurst, 2003). The compressor will construct a dictionary for the common sequences found within the information and then match each sequence to a shorter compressed representation or a key code.

Two types of dictionary, namely a running dictionary which based on the compression algorithm used or a pre-defined dictionary that can be used for bulk compression. In bulk compression, the decompressor must use an identical dictionary which is used during compression and bulk compression is known to achieve higher compression ratio. However, the data dictionary requires larger memory allocation and the dictionaries at both the compressor and decompressor sides have to be fully synchronized.

4.3.1 Payload compression schemes

Apart from packet and header compression, two payload compression schemes have been proposed by other researchers, which are Adaptive Compression Environment (ACE) system and Adaptive Online Compression (AdOC) algorithm.

4.3.1.1 Adaptive Compression Environment (ACE)

Adaptive Compression Environment (ACE) intercepts program communication and applies on-the-fly compression (Krintz & Sucu, 2006). On-the-fly or online compression is mandatory for those real-time interactive applications. ACE is able to adapt to the changes in resource performance and network technology, thus the benefits from using ACE become apparent when the underlying communication performance varies or the network technology changes as in mobile communication network. ACE employs an efficient and accurate forecasting toolkit, which is known as Network Weather Service (NWS) to predict and determine whether applying compression will be profitable based on underlying resource performance.

Short-term forecasts of compression ratio, compressed and uncompressed transfer time is made by NWS using a series of estimation techniques, together with its own internal models that estimate compression performance and changes in data compressibility.After that, based on the end-to-end path information obtained by NWS, ACE will select between several widely used compression techniques, which include bzip, zlib and LZO to perform the transparent compression at TCP socket level. ACE compresses data in 32KB blocks and a 4-byte header is appended to each block to indicate the block size and compression technique used. It is proven to improve transfer performance by 8-93 percent over commonly used compression algorithm (Krintz & Sucu, 2006).

However, ACE may introduce computation overheads due to massive amount of computation are needed during the prediction process. Besides, problem like prediction error which will lead to inaccurate decision may occur and large compression time cost of the compression algorithm such as bzip may impose additional delays. Thus, ACE may not be suitable and may impose additional delays over satellite network.

4.3.1.2 Adaptive Online Compression (AdOC)

This work proposed a general purpose portable application layer compression algorithm known as AdOC. AdOC is an adaptive online compression algorithm suited for any application data transfer and it automatically adapts the level of compression to the speed of the network (Jeannot et al., 2002). Multithreading and First-In-First-Out (FIFO) data buffer are two important features of this algorithm.

In this algorithm, the sender consists of two threads, namely compression thread and communication thread. Compression thread is used to read and compress the data, while communication thread is responsible to send the data. A FIFO data buffer is created to store the data prior to transmission. The compression thread will write the data into the FIFO data buffer, while the communication thread will retrieve the data from it. Thus, the compression level used in the process of compression is depending on the size of the FIFO queue.

To completely eliminate the overhead encountered when data cannot be compressed, AdOC algorithm compresses data into smaller and independent chunks. This made AdOC less reactive to short term changes in bandwidth, but keeping the same compression level for long runs of data also improves the compression ratio (Jeannot et al., 2002). However, too small chunks of data will simply caused overhead of FIFO queue, hence, the size of data chunks need to be determined appropriately. Since AdOC algorithm compresses data into smaller and independent chunks, network load may be increased and network congestion may occur when works under satellite network.

5. Proposed real-time adaptive packet compression scheme

An overview of the proposed real-time adaptive packet compression scheme, with the highlighting of its main concept and properties, is provided in this section. The block diagram of the proposed compression scheme, together with the explanation of each stage involved is also presented.

5.1 Concept of the proposed scheme

Concept of the proposed real-time adaptive packet compression scheme in satellite network topology is shown in Figure 10 below. As stated earlier, the main objective of this research study is to overcome the limitation and constraints of satellite communication link, which are high latency and low bandwidth, therefore the performance of the satellite link has become the main consideration in the proposed scheme. The proposed approach will focus only on the high latency satellite link area, where the proposed scheme will be implemented in both gateway A and gateway B. Both gateways will act as either compressor or decompressor as the communication channel between gateway A and gateway B is a duplex link.

In the proposed compression scheme, the concept of virtual channel is adopted to increase network performance and reliability, simplify network architecture, and also improve network services. Virtual channel is a channel designation which differs from the actual communication channel and it is a dedicated path designed specifically for both sender and receiver only. Since packet header compression is employed in the proposed scheme, thus this concept is mandatory to facilitate data transmission over the link. The duplex link between gateway A and gateway B in Figure 10 will act as the virtual channel, where the rules of data transmission and the data format used are agreed by both gateways.

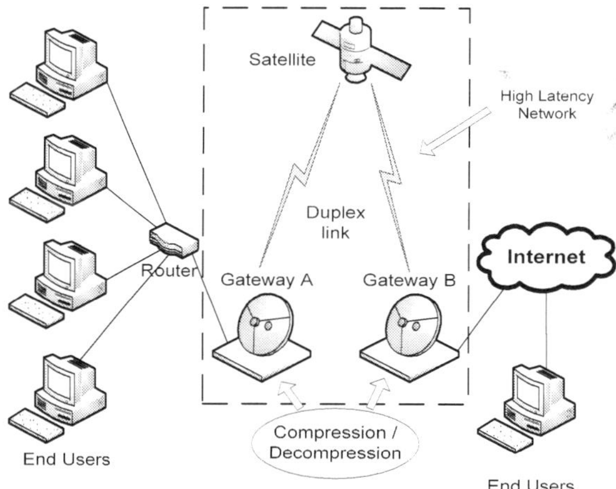

Fig. 10. Concept of the proposed compression scheme.

The flow of data transmission between both gateways is as discussed in the following. When the transmitted data packets arrive at gateway A, the packets will undergo compression prior to transmission over the virtual channel. When the compressed data packets reach gateway B, the compressed packets will first undergo decompression before being transmitted to the end user.

Apart from that, adaptive packet compression is mandatory due to the adoption of block compression in the proposed scheme. Although block compression helps to increase the compression ratio, however, it has its downside too. Block compression might impose additional delay when the compression buffer is filled in a slow rate due to lack of network traffic and a fast response is needed. This will further reduce the user experience of VSAT satellite network. Therefore, to avoid this, packet blocks are compressed adaptively when any of the predefined conditions is reached, which will be discussed in details in the following section.

5.2 Strength of the proposed scheme

The proposed real-time adaptive packet compression scheme has several important properties as discussed in the following. Firstly, the proposed scheme is accommodating all incoming packets. To fully exploit the positive effect of compression, the proposed scheme is not restricted to specific packet flow only but is applied to all incoming packets from numerous source hosts and sites. One unique feature of the proposed scheme is the adoption of virtual channel concept, which has not been used in other reviewed schemes. This concept simplifies packet routing and makes data transmission more efficient, especially when packet compression is employed. In the proposed scheme, to facilitate packet transmission over the communication channel, a peer-to-peer synchronized virtual channel is established between the sender (compressor) and receiver (decompressor). Moreover, another important feature, block compression approach is also introduced. Block compression exploits similarities of consecutive packets in the flow and compression is performed on an aggregated set of packets (a block) to further improve the compression ratio and increase the effective bandwidth.

Apart from that, both packet header and payload are being compressed in the proposed scheme. In many services and applications such as Voice over IP, interactive games and messaging, the payload of the packets is almost of the same size or even smaller than the header (Effnet, 2004). Since the header fields remain almost constant between consecutive packets of the same packet stream, therefore it is possible to compress those headers, providing more than 90% (Effnet, 2004) saving in many cases. This helps to save bandwidth and the expensive resources can be used efficiently. In addition to header compression, payload compression also introduces significant benefit in increasing the effective bandwidth. Payload compression compresses the data portion of the transmission and it uses compression algorithms to identify relatively short byte sequences that are repeated frequently over time. Payload compression provides a significant saving in overall packet size especially for packets with large data portions.

In addition, adaptive compression is employed in the proposed scheme. Network packets are compressed adaptively and selectively in the proposed scheme to exploit the positive effect of block compression while avoiding the negative effect. To avoid greater delay imposed by block compression, the set of aggregated packets (block of packets) in the compression buffer is compressed adaptively based on certain conditions. If either one of the conditions is fulfilled, the compression buffer is compressed. Else, the compression buffer will not be compressed. By combining all the features listed above, the performance of the proposed scheme will be greatly improved over other reviewed schemes.

5.3 Overview of the proposed scheme

Figure 11 below demonstrates the main components of the proposed real-time adaptive packet compression scheme. The compression scheme made up of a source node (Gateway A) which acts as the compressor and a destination node (Gateway B) which is the decompressor. A peer-to-peer synchronized virtual channel, which acts as a dedicated path, will be established between Gateway A and Gateway B. With the presence of virtual channel, packet header compression techniques can be performed on all network packets.

Data transmission between Gateway A and Gateway B can be divided into three major stages, which are compression stage, transmission stage and decompression stage. Compression stage takes place in Gateway A, transmission stage in the virtual channel while the decompression stage will be carried out in Gateway B. Every data transmission from Gateway A to Gateway B will undergo these three stages.

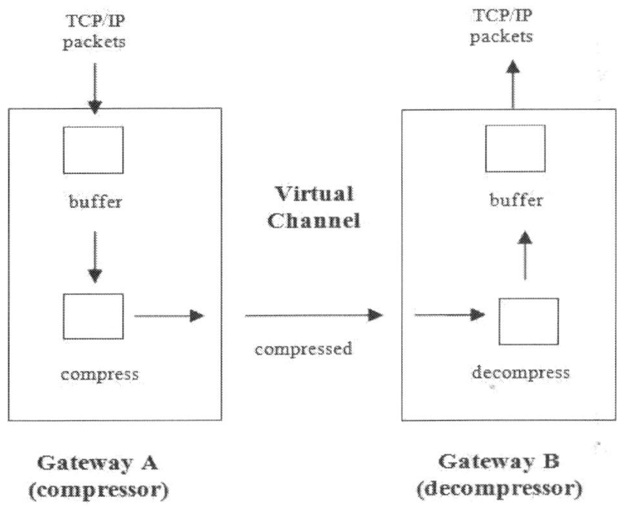

Fig. 11. Main components of the proposed compression scheme.

5.3.1 Compression stage

Once the incoming packets reach the Gateway A, the packets will be stored inside a buffer. This buffer is also known as compression buffer, as it is used for block compression, which will be discussed in details in the following section. Generally, in block compression, packets will be aggregated into a block prior to compression. The buffer size is depending on the maximum number of packet which is allowed to be aggregated.

Block compression is employed to increase the compression ratio and reduce the network load. The compression ratio increases with the buffer size, which means that the larger the buffer, the better the compression ratio, as more packets can be aggregated. However, block compression may lead to higher packet delays due to the waiting time in the buffer and also the compression processing time. The packet delay time is expected to increase with the number of packet to be aggregated. Thus, larger buffer will have higher compression processing latency and also higher packet drops. Therefore, a trade off point is mandatory. Once the whole compression buffer fills up, it will be transferred to the compress module to undergo compression. The compression buffer will be compressed via a well known compression library known as zlib compression library (Roelofs et al., 2010). One apparent drawback of this scheme with block compression is a possible delay observed when the compression buffer is filled in a slow rate due to lack of network traffic and a fast response is needed. To address this shortcoming, the proposed scheme will compress the compression buffer adaptively whenever any of the following conditions are met:

a.   The compression buffer reaches its predefined limit or has filled up.
b.   A certain time threshold has been exceeded from the time the first packet being stored in the buffer and the buffer contains at least one packet.

After the process of compression, the compressed block will now enter the transmission stage.

5.3.2 Transmission stage

In this stage, the compressed block will be transmitted over the communication link, which is a virtual channel in this scheme, to Gateway B. The compressed block will transit from transmission stage to decompression stage when it reaches the Gateway B.

5.3.3 Decompression stage

The compressed block will be directly transferred to the decompress module once it reaches Gateway B. Decompression will then be performed on it to restore its original form. The original block of packets will be divided into individual packets according to the original size of each combined packet. After that, these individual packets are stored in the decompression buffer while waiting to be transmitted to the corresponding end user or destination node.

5.4 Block compression

Block compression exploits similarities of consecutive packets in the flow, as a specific number of packets are aggregated into a block before undergo compression. Due to the correlation of packets inside the packet stream, the compression ratio is greatly improved. Besides, block compression helps to reduce the heavy network load and avoid network congestion. This is because it reduces the number of packets needed to be transmitted over the communication link by encapsulating a significant number of individual packets into a large packet (block).

An example of block compression, where four network packets are collected in a compression buffer before being compressed and transmitted to the receiver, is shown in Figure 12. As mentioned earlier, one of the shortcoming of block compression is it may potentially add great packet delays, as the packets do not immediately be transmitted but instead stored in the compression buffer. This packet delay time is expected to increase with the number of packet to be combined.

For example, Table 1 below shows the total number of accumulated transmitted packet in 5 unit time for a high latency network with compression scheme (HLNCS) and a high latency network without compression scheme (HLN). Suppose that the number of packet to be encapsulated for the high latency network with compression scheme is 10.

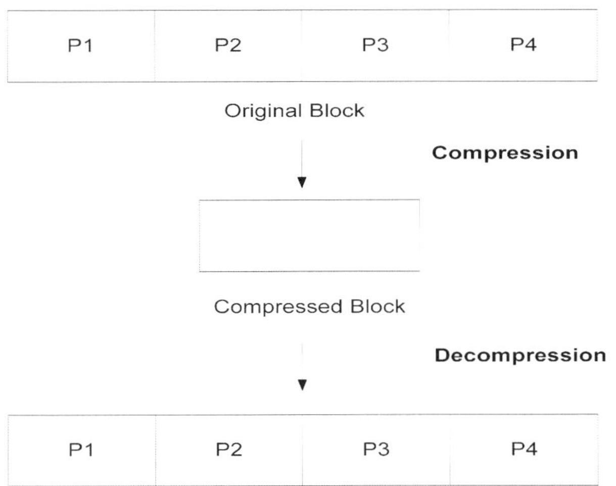

Fig. 12. Block compression.

HLN		HLNCS	
Time	No. of packet transmitted	Time	No. of packet transmitted
1st	1	1st	0
2nd	2	2nd	0
3rd	3	3rd	0
4th	4	4th	0
5th	5	5th	10
Total	5	Total	10

Table 1. No. of transmitted packet for HLN & HLNCS.

Note that for HLN, there is no delay in transmitting the packet in each unit time and 5 packets are sent after 5 unit time, while for HLNCS, there is 4 unit time delay and 10 packets are transmitted at 5 unit time. Due to the waiting time in the compression buffer and the compression processing time, packet transmission is delayed. However, the total number of packet transmitted is almost double even though there is a small delay initially. Thus, with tolerable delay, block compression allows more packets to be sent at one time. A trade off value between the packet delay and number of packets to be combined needs to be determined.

6. Results & discussions

In this section, the proposed real-time adaptive packet compression scheme is evaluated and validated by simulations. Two important performance metrics of the scheme, which are packet drop rate and throughput of data transmission, are evaluated, as these two metrics are representing the Quality of Service of satellite link. The performance criteria can be defined as the following. Packet drop rate is the ratio between the total amount of packet loss due to buffer overhead (congestion) and transmission errors, and the total amount of packets being transmitted successfully, in percentage. Throughput is the ratio between the total amount of packets successfully delivered to the receiver and the time of the connection (2000 seconds). A discrete event network simulator known as ns-2 (VINT Project, 1995) has been used in building the simulation model to realize a simulative framework for studying and evaluating the performance of the proposed real-time adaptive packet compression scheme over high latency satellite network environment.

6.1 Simulation setup

This section describes the experimental environment used to present the characteristics and effectiveness of the proposed scheme. Figure 13 below depicts the simulation network topology, where *n* users are connected to a source node through wired links and the source node is connected to the destination node via the high latency satellite communication link. Each wired link presents a capacity of 10 Mbit/s and a propagation delay of 1 ms. The proposed real-time adaptive packet compression scheme is implemented at both source and destination nodes. Different values of number of user and various satellite link characteristics are simulated to monitor the impact of the proposed scheme over satellite link. TCP continuous traffic flows are used throughout the simulations. All users transmit packets simultaneously to the destination node via the source node and each simulation is run for 2000 seconds.

The effectiveness of the proposed scheme is evaluated by comparing the performance metrics (packet drop rate and throughput of data transmission) of two scenarios: simulation running with proposed scheme and simulation running without the proposed scheme. For the scenario with proposed scheme, packet is compressed in the source node before transmitting over the satellite link and is decompressed when it reaches the destination node. For the scenario without the proposed scheme, normal data transmission is carried out. The packet trace data used throughout the simulations are captured from the research labs in University Malaysia Sarawak (UNIMAS), which composed of normal day-to-day traffic, typical for research purposes. The traces are taken by using a traffic capture utility known as Wireshark (Wireshark Foundation, 1998).

As shown in the Table 2 below, different simulation scenarios are used to evaluate the proposed scheme. Two scenarios, low bandwidth and high bandwidth, are simulated. In each scenario, five different number of user are used to vary the congestion rate of the satellite link, so that the impact of the proposed scheme on link with different congestion values can be examined. The compression rate used in the compression process is also varied for each value of number of user used, as depicted in Table 2. Compression rate is the size of the compression buffer for block compression. For example, compression rate with value 0 means no compression, compression rate with value 1 means packet by packet compression, compression rate with value 5 means 5 packets are to be aggregated in the compression buffer prior to compression, and so on.

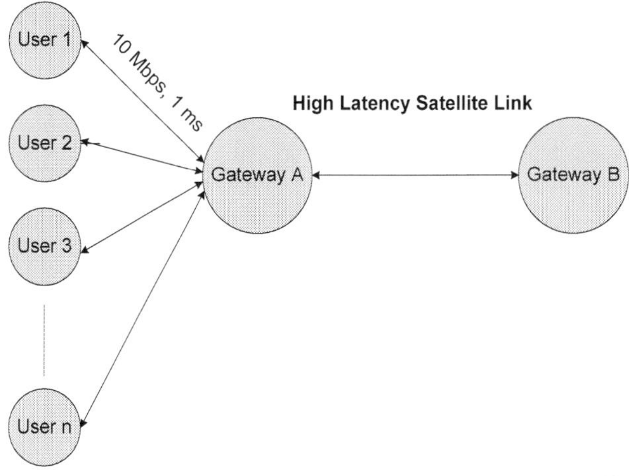

Fig. 13. Simulation topology.

Scenario	Satellite link characteristic	User	Compression rate
1 (low bandwidth)	Uplink bandwidth: 64kbps Downlink bandwith: 256kbps Round-trip delay: 644ms	5, 15, 25	0, 1, 5 – 1000 (with step of 5)
		35, 45, 55	0, 1, 5 – 2000 (with step of 5)
2 (high bandwidth)	Uplink bandwidth: 1024kbps Downlink bandwith: 2048kbps Round-trip delay: 644ms	5, 15, 25	0, 1, 5 – 1000 (with step of 5)
		35, 45, 55	0, 1, 5 – 2000 (with step of 5)

Table 2. Simulation scenarios.

6.2 Performance analysis

As discussed in previous section, block compression is employed in the proposed scheme and different sizes of the compression buffer (compression rate) are used in the simulation studies. Block compression helps to improve the packet throughput as more packets can be transmitted over the communication channel at the same time. However, it may lead to higher packet drop rate as the whole packet block will be discarded when it encountered errors or when it is lost in the middle of transmission. This condition is getting worse when a high compression rate is used. Thus, an appropriate compression rate is crucial in achieving a high packet throughput with acceptable packet drop rate. The tolerable value for packet drop rate is depending solely on the application requirements.

From the simulation results, compression rate which yields the highest packet throughput given that the packet drop rate is less than 5%, 10% and 15%, is selected. Thus, the results can be divided into three cases. Case 1 considers packet drop rate not more than 5%, Case 2 limits the packet drop rate to 10% and packet drop rate less than 15% is considered in Case 3. A communication link with packet drop rate more than 15% is considered as a bad performance link even though the throughput obtained is very high. Therefore, packet drop rate more than 15% is beyond the consideration in this work. The results are presented in the following section.

6.2.1 Best compression rate distribution

Figure 14 & 15 below shows the distribution of best compression rate over the congestion rate for Case 1,2 and 3 in Scenario 1 & 2. Best compression rate is the compression rate that yields the highest throughput, with the condition that its corresponding packet drop rate does not exceed the limit in each case. Notice that in both scenarios, the best compression rate increases with the congestion rate. This means that a larger compression buffer, which can accommodate more packets, is favoured to obtain a higher performance when the link is getting more and more congested. In Scenario 1, due to the packet drop rate constraints that limits the highest throughput that can be achieved, the best compression rate line of Case 1 is slightly lower than the line of Case 2 & 3, while Case 2 & 3 both achieve similar results (overlapped lines). In Scenario 2, all three cases favour the same compression rates.

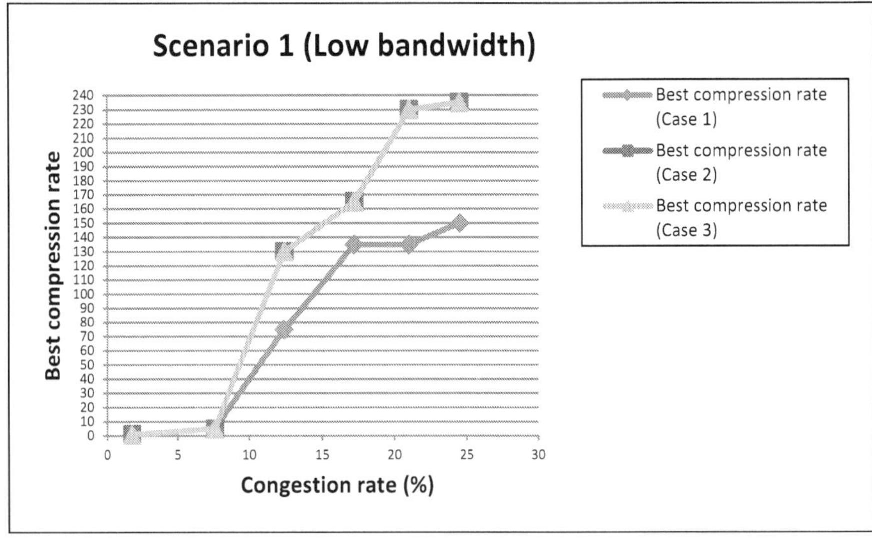

Fig. 14. The best compression rate for each case in Scenario 1.

As shown in the figures, low bandwidth scenario requires higher compression rates (1 - 235) while high bandwidth scenario requires lower compression rates (1 - 15). This shows that the proposed scheme performs better in low bandwidth scenario compared to high bandwidth scenario. This is because high bandwidth scenario has sufficient bandwidth to accommodate heavy flows of traffic, thus compression might not be needed, while in the case of low bandwidth, compression is mandatory as bandwidth limitation problem will cause the communication link to be severely congested.

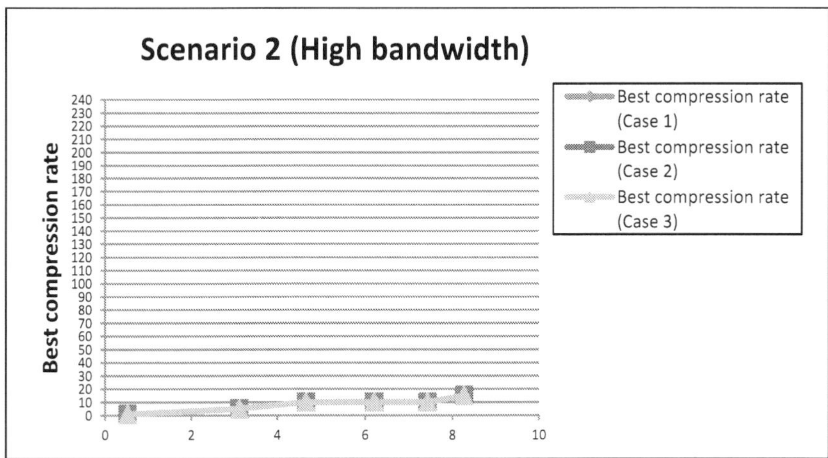

Fig. 15. The best compression rate for each case in Scenario 2.

6.2.2 Packet drop rate distribution

Figure 16 & 17 above shows the distribution of packet drop rate over the congestion rate for the simulation running without the proposed scheme and simulation running with the proposed scheme (Case 1, 2 and 3) in Scenario 1 & 2. Notice that in both scenarios, the packet drop rate of simulation without compression increases with the congestion rate. This means that communication link with higher congestion value has higher packet drops. With the adoption of the proposed scheme, block compression reduces the heavy network load and hence avoiding network congestion. Thus, the packet drop rate can be reduced significantly as no packet being dropped due to buffer overhead at the router.

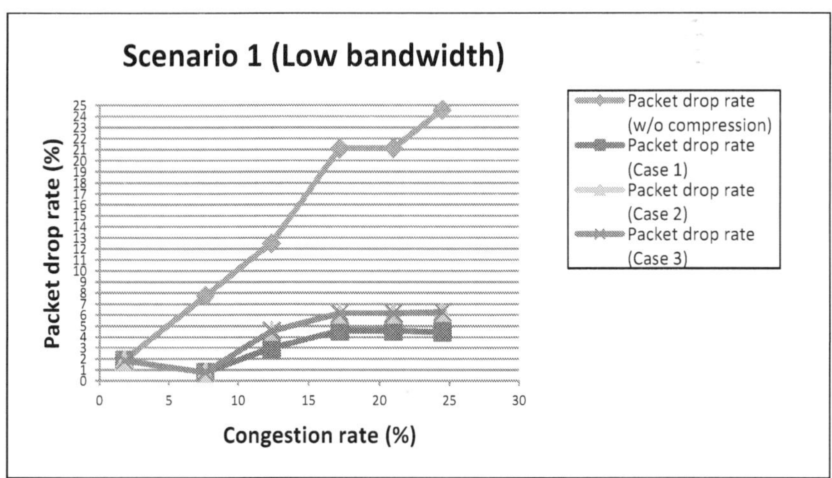

Fig. 16. The packet drop rate distribution in Scenario 1.

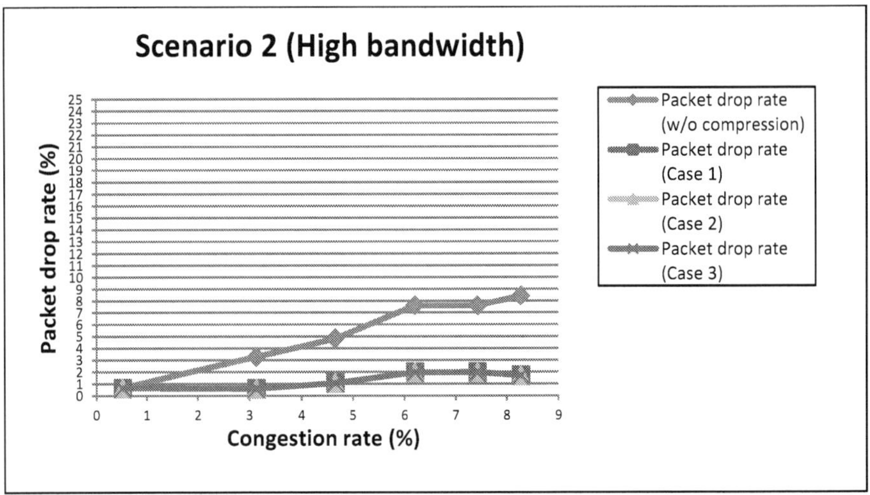

Fig. 17. The packet drop rate distribution in Scenario 2.

The proposed scheme succeeds in reducing the packet drop rate by 1 – 90 percent in Scenario 1 and 1 – 82 percent in Scenario 2. In Scenario 1, due to the packet drop rate constraint of 5%, the packet drop rate line of Case 1 is slightly lower than the line of Case 2 & 3, which with the limits of 10% & 15%. Since the best compression rates for Case 1, 2 & 3 are similar as illustrated in Figure 15, thus the corresponding packet drop rate values of these three cases are the same too.

6.2.3 Packet throughput distribution

Figure 18 & 19 below shows the distribution of packet throughput over the congestion rate for the simulation running without the proposed scheme and simulation running with the proposed scheme (Case 1, 2 and 3) in Scenario 1 & 2. The throughput for simulation without compression decrease with the congestion rate in both scenarios. The more congested the link, the more packets being dropped due to buffer overhead at the router, hence the lower the throughput. As shown in Figure 18 & 19, the proposed scheme succeeds in improving the throughput by 8 – 175 percent in Scenario 1 and 5 – 62 percent in Scenario 2. This is because by applying block compression, more packets can be transmitted over the communication link at one time, hence the throughput can be greatly improved.

Notice that the improvement of packet throughput in Scenario 1 is better than in Scenario 2. This also suggests that the proposed scheme is performing much more better in a low bandwidth scenario compared to a high bandwidth scenario. This is because compression might not be necessary in high bandiwdth scenario, as there is no bandwidth limitation problem and sufficient bandwidth is provided to accommodate heavy flows. In contrast, applications are competing for the low and limited bandwidth when there are heavy flows in a low bandwidth scenario, thus, compression is required to further improve the network performance.

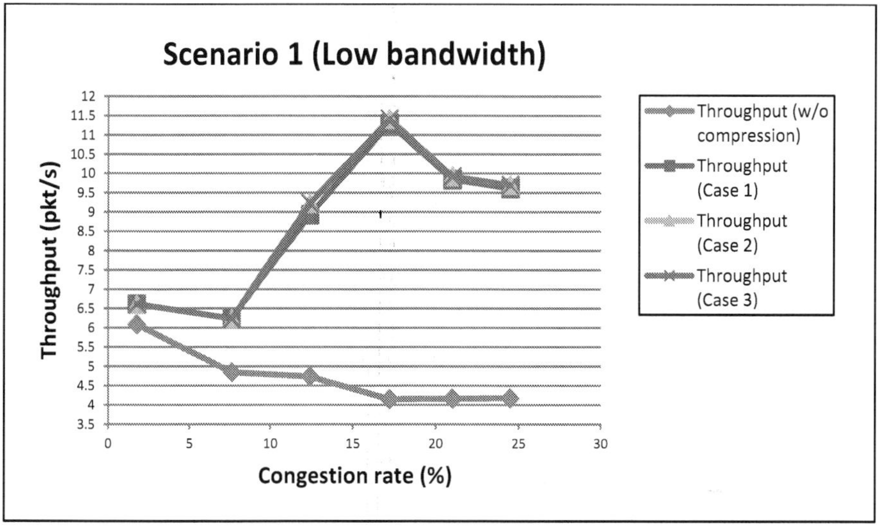

Fig. 18. The packet throughput distribution in Scenario 1.

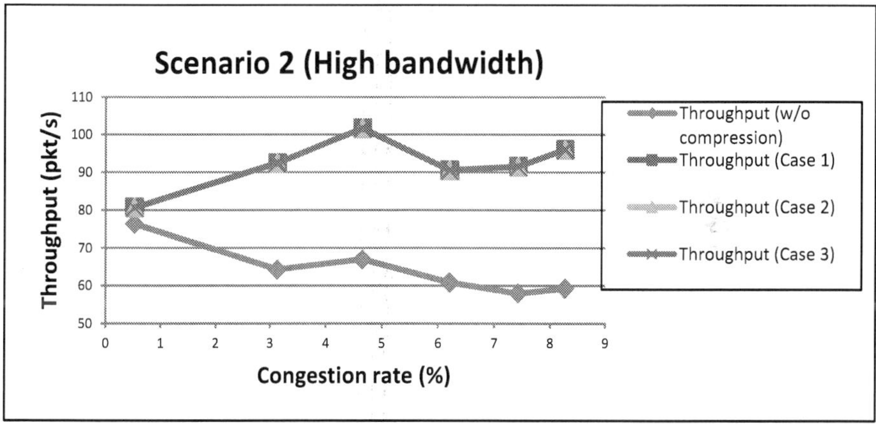

Fig. 19. The packet throughput distribution in Scenario 2.

7. Conclusion

In this chapter, a real-time adaptive packet compression scheme for bandwidth limited high latency satellite communication network is presented. The bandwidth limitation problem of high latency satellite network has lead to several crucial network issues, as more and more applications require higher bandwidth allocation. The proposed scheme is intended to improve Quality of Service of real-time interactive applications by increasing the effective bandwidth usage of satellite network. Besides employing both header and payload compression to achieve maximum bandwidth optimization, this scheme facilitates communication and reduces network processing complication by establishing a virtual channel between sender and receiver. As discussed earlier, virtual channel is a channel designation which differs from the actual communication channel and it is a dedicated path designed specifically for both sender and receiver only. Thus this concept is mandatory to facilitate data transmission over the link, as packet header compression is employed. Block compression is also adopted in the compression scheme to improve the compression ratio and reduce the network load.

To evaluate the performance and effectiveness of the proposed scheme, extensive simulations have been conducted using captured TCP traffic. The proposed scheme is evaluated under two main scenarios: low bandwidth and high bandwidth. Simulation results show that the proposed scheme succeeds in reducing the packet drop rate and improving the packet throughput significantly in both low and high bandwidth scenarios, as shown in Table 3.

Scenario	Improvement Percentage (%)	
	Packet drop rate	Packet throughput
Low bandwidth	Up to 90	Up to 175
High bandwidth	Up to 82	Up to 62

Table 3. Improvement percentage on packet drop rate and packet throughput.

Hence, it is proven that through the introduction of this scheme, the Quality of Service of real-time interactive applications over high latency satellite network can be greatly improved as the main concern of satellite network which is low and limited bandwidth is now not an issue anymore. Real-time interactive applications and software, which have high bandwidth demand, will now gain good user experience and satisfaction over satellite network.

8. Acknowledgement

This research work was funded by Fundamental Research Grant Scheme – FRGS/02(16)/737/2010(23), Universiti Malaysia Sarawak (UNIMAS). The authors would like to thank Faculty of Computer Science and Information Technology, UNIMAS for providing useful equipments and facilities. The authors would also like to thank the anonymous reviewers for their comments which helped in improving the quality of this chapter.

9. References

Chen, S.; Ranjan, S. & Nucci, A. (2008). IPzip: A Stream-aware IP Compression Algorithm, *Proceedings of IEEE Data Compression Conference,* pp. 182-191, ISBN 978-0-7695-3121-2, Snowbird, Utah, USA, March 25-27, 2008

Comsys. (2008). VSAT statistics from COMSYS, 07.04.2011, Available from http://www.comsys.co.uk/wvr_stat.htm

DotNetNuke Corporation. (2010). Sat.com Satellite Information Site, 12.02.2011, Available
    from http://www.satelite.com/

Effnet. (2004). The Concept of Robust Header Compression, ROHC, 20.02.2011, Available
    from http://www.effnet.com/sites/effnet/pdf/uk/
    Whitepaper_Robust_Header_Compression.pdf

Hart, D. (1997). Satellite Communications, 14.02.2011, Available from
    http://www1.cse.wustl.edu/~jain/cis788-97/ftp/satellite_nets.pdf

Jacobson, V. (1990). Compression TCP/IP for Low-Speed Serial Link, *RFC 1144*, 1990

JCP-Consult. (2008). JCP-C RoHC Headers Compression Protocol Stack, pp. 1-9, Cesson-
    Sevigne, France

Jeannot, E.; Knutsson, B. & Bjorkman, M. (2002). Adaptive Online Data Compression,
    *Proceedings of 11th IEEE International Symposium on High Performance Distributed
    Computing*, pp. 379, ISBN 0-7695-1686-6, Edinburgh, Scotland, July 24-26, 2002

Krintz, C. & Sucu, S. (2006). Adaptive On-The-Fly Compression, *IEEE Transaction on Parallel
    and Distributed Systems*, Vol.17, No. 1, January, 2006, pp. 15, ISSN 1045-9219

Matias, Y. & Refua, R. (2005). Delayed-Dictionary Compression for Packet networks,
    *Proceedings of 24th Annual Joint Conference of the IEEE Computer and Communications
    Societies*, pp. 1443-1454, ISBN 0-7803-8968-9, Miami, Florida, USA, March 13-17,
    2005

MindBranch. (2011). World VSAT markets (raw data spreadsheet included): Industry
    Report, 07.04.2011, Available from
    http://www.mindbranch.com/listing/product/R1-4772.html

Mitra, M. (2005). *Satellite Communication*, Prentice-Hall of India Private Ltd, ISBN 978-81-203-
    2786-3, New Delhi, India

Naidu, D. & Tapadiya, R. (2009). Implementation of Header Compression in 3GPP LTE, *6th
    International Conference on Information Technology: New Generations*, ISBN 978-0-7695-
    3596-8, Las Vegas, Nevada, April 27-29, 2009

Pu, I. (2006). *Fundamental Data Compression*, Butterworth-Heinemann, ISBN 978-0-7506-6310-
    6, Burlington, Massachusetts

Richharia, M. (1999). *Satellite Communication Systems: Design Principles* (2nd Ed.), Macmillan
    Press Ltd, ISBN 0-333-74722-4, London, England

Roelofs, G.; Gailly, J. & Adler, M. (2010). zlib Home Site, 20.02.2010, Available from
    http://www.zlib.net/

Shimamura, M.; Ikenaga, T. & Tsuru, M. (2009). Compressing Packets Adaptively Inside
    Networks, *Proceedings of 9th Annual International Symposium on Applications and the
    Internet*, pp. 92, ISBN 978-1-4244-4776-3, Seattle, USA, July 20-24, 2009

Sun, Z. (2005). *Satellite Networking: Principles and Protocols*, John Wiley & Sons Ltd, ISBN 978-
    0-470-87027-3, West Sussex, England

Suryavanshi, V.; Nosratinia, A. & Vedantham, R. (2004). Resilient Packet Header
    Compression through Coding, *Proceedings of Global Telecommunications Conference*,
    pp. 1635, ISBN 0-7803-8794-5, Dallas, Texas, USA, November 29 – December 3, 2004

Tan, L.; Lau, S. & Tan, C. (2010). Enhanced Compression Scheme for High Latency
    Networks to Improve Quality of Service of Real-Time Applications, *Proceedings of
    8th Asia-Pacific Symposium on Information and Telecommunication Technologies*, pp. 1-6,
    ISBN 978-1-4244-6413-5, Kuching, Sarawak, Malaysia, June 15-18, 2010

Taylor, D.E.;   Herkersdorf, A.; Doring, A. & Dittmann, G. (2005). Robust Header Compression (ROHC) In Next-Generation Network Processors, *IEEE/ACM Transactions on Networking*, Vol. 13, No.4, August, 2005, pp. 755-768, ISSN 1063-6692

Telekom   Malaysia   Berhad.   (2011).   VSAT,   07.04.2011,   Available   from http://202.71.108.103/business/corporate-government/data-services/vsat/faqs.asp

TopBits.com. (2011). VSAT, 07.04.2011, Available from http://www.tech-faq.com/vsat.html

Tye, C.S. & Fairhurst, D.G. (2003). A Review of IP Packet Compression Techniques, *PGNet*, ISBN 1-9025-6009-4, Liverpool, UK, June, 2003

VINT   Project   (1995).   The   network   simulator – ns-2,   11.11.2009,   Available   from http://www.isi.edu/nsnam/ns/

Wireshark   Foundation   (1998).   Wireshark.   Go   deep,   11.12.2009,   Available   from http://www.wireshark.org/

Part 4

Hybrid Satellite-Terrestrial Networks

4

Multicast Security and Reliable Transport of Rekey Messages over Hybrid Satellite/Terrestrial Networks

Franco Tommasi, Elena Scialpi and Antonio De Rubertis
University of Salento - Department of Engineering for Innovation
Lecce (Italy)

1. Introduction

Security problems in satellite environments are one of the obstacles to the widespread deployment of satellite IP multicast and, more generally, of satellite multimedia applications (Cruickshank et al., 1998).

By satellite environments we refer to networks where the satellite plays an essential role. e.g. those where it is used to multicast IP packets to many nodes of a terrestrial network. We also speak of "Hybrid Satellite/Terrestrial networks" in such cases.

The broadcast nature of satellites makes eavesdropping and active intrusion much easier than in terrestrial fixed or mobile networks. A further issue is specific to multicast: the number of members in a multicast group can be very large and, even worse, can change very dynamically. While the process of performing and securing key management for unicast connections is well understood (Harkins & Carrel, 1998), (Maughan et al., 1998), (Orman, 1998), multicast security is still an open field (see par. 2). Protocols that manage the process of distributing keys in a multicast environment are under development (see par. 2.3 and 2.4).

Access to the encryption key is controlled by a group key management system, which is responsible for sending the encryption key to authorized new users and for performing multicast group rekeying whenever the key changes. Specifically, a group key management system is said to implement two types of access control: backward access control and forward access control. If the system changes the encryption key after a new user joins, the new user will not be able to decrypt past group communications; this is called backward access control. Similarly, if the system rekeys after a current user leaves, or is expelled from the system, the departed user will not be able to access future group communications; this is called forward access control.

Many group key management solutions (see par. 2.2, (Jokela, 2006) (Mah, 2004)) have been proposed and a number of classifications of the available approaches can be found in the current literature (Dondeti et al., 1999), (Rafaeli & Hutchison, 2003), (Eskicioglu, 2003). Moreover, security mechanisms regarding satellite networks have been investigated in (Howarth et al., 2004), (Noubir & Allmen, 1999) and (Arslan & Alagöz, 2006).

Group key management protocols can be categorized as following:

- *Centralized architectures.* A single entity, a GC (Group Controller), is employed for controlling the whole group, hence a group key management protocol seeks to minimize storage requirements, computational power on both client and server sides and bandwidth utilization.

- *Decentralized architectures*. The management of a large group is divided among subgroup managers, trying to reduce the problems arising from concentrating the work in a single place.

- *Distributed architectures*. There is no explicit manager and the members themselves do the key generation. All members can perform access control and the generation of the key can be contributory, meaning that all members contribute some information to generate the group key.

Rekey protocols should use a scalable Group Key Management Algorithm (GKMA) to send the minimum possible number of keys in a rekey message. LKH (see par. 2.3), OFT (Balenson et al., 2000), Subset difference based schemes (Lotspiech et al., 2001) are examples of GKMA. Regardless of the chosen approach, rekey messages are generally frequent and their reception must be guaranteed in order for the multicast group members to avoid multicast services interruptions.

RFC 4046 (Baugher et al., 2005) describes a Group Key Management Architecture and proposes three classes of solutions for reliably sending keys to the multicast group members:

- repeatedly transmit the rekey message;

- use FEC for encoding rekey packets (with NACKs as feedback) (Yang et al., 2001);

- use an existing reliable multicast protocol/infrastructure (possibly profiting in a mixed way from the above solutions).

Up to now, not much work has been dedicated to the use of reliable multicast transports for rekey messages. In most cases ((Wong & Lam, 2000) (Zhang et al., 2003)) FEC (Rizzo, 1997) has been used to improve the reliability.

RFC 4046 also identifies the requirements a protocol for key transmission/rekeying must satisfy:

- *Reliability*. Every user must receive all of its (encrypted) new keys, no matter how large the group size.

- *Soft real-time*. It is required that the delivery of new keys to all users be finished with a high probability before the start of the next rekeying.

- *Scalability*. The processing and bandwidth requirements of the key server and those of each user should not increase much with the group size so that a single server is able to support a large group.

Moreover, multicast key distribution must take care of the "feedback implosion" problem (see par. 2.2.4 and (Baugher et al., 2005) resulting from NACKs or ACKs sent as feedback.

Satellite networks may intrinsically offer a serious alternative to terrestrial networks solutions in that they can enable reliable multicast techniques to scale to large group of receivers. Such advantage is an effect of their intrinsic properties such as: high bandwidth availability, their broadcast nature and the reduced occurrence of congestion between sender and receivers as compared to terrestrial networks.

With these considerations in mind, we focused our attention on the following protocols for the multicast reliable transmission of encryption keys: Pragmatic General Multicast (PGM) (see par. 3.1), NACK-Oriented Reliable Multicast (NORM) (see par. 3.2) and our SRDP-Sign (see par. 3.3). PGM was chosen for being an interesting IETF experimental protocol. While not yet a standard, it has been implemented in some networking devices (such as Cisco routers) and operating systems including MS Windows XP. NORM was chosen because RFC 4046 quotes it as a well-suited protocol for reliable multicast of rekey messages.

In the following, paragraph 2 will detail the state of the art on the subject of multicast security with a particular attention to the solutions based on a centralized approach, paragraph 3 will discuss some reliable multicast protocols with an interest in their utilization in satellite networks. Paragraph 4 will present the preliminary results of some tests we conducted with the aim of evaluating the performances of above listed reliable multicast protocols. They have been tested on a hybrid satellite/terrestrial network in the specific case of transmission/rekeying of keys for a multicast security environment.

2. Multicast security

The original conception of an IP network was aimed at the exchange of information between two nodes. However, very soon the popularity of the Internet gave rise to a number of applications for which a better model would be desirable. Such applications would benefit from a network direct support to the delivery of the same packets from one source to many destinations. Some of them are today's killer applications, e.g. IPTV. The need for multiple unicast connections implied by the basic model made them simply not scalable enough within the original rules.

Around 1989, to address such problem the introduction of a new functionality was proposed for IP networks: IP multicast (Deering, 1991) (Deering, 1989). As a result of it, an host wishing to send the same packet to many hosts at the same time was allowed to output that single packet on its network interface, leaving to the network's routers the burden of duplicating it wherever required. As an extreme example, a packet intended for a number of hosts on a distant LAN would travel alone until the last router which would replicate it at the last hop for as many hosts as needed.

The positive effect of such approach can be perceived, increasingly with the number of the multicast group members, both on the conservation of computational resources of the sending machine and in the (potentially huge) savings of bandwidth resources in the network.

The idea required the introduction of a special class of IP addresses (Class D, from 224.0.0.0 to 239.255.255.255) each of them representing a "multicast group".

The essential protocol for managing the multicast group membership is IGMP (Internet Group Multicast Protocol) (Deering, 1989). It works without problems in a network where all the routers support it. When support is spotty, more complex techniques are required (Semeria & Maufe, 1997).

Although IP Multicast would be the ideal technique for many important applications (e.g. to distribute real-time video on the Internet) for many well-known reasons it is not globally supported on the Internet (Diot et al., 2000). There are indeed many ISPs supporting IP multicast in their AS (Autonomous Systems) and multicast peering agreements are frequent among ISPs but even then the common user isn't left the faculty to send multicast traffic to other users in the same AS. Clearly this ability is regarded as a primary asset within an ISP network and acquiring it (when available) can be subjected to substantial fees. Many methods to overcome this limitation have been proposed (Eriksson, 1994) (*MBONED*, 2011) (Sardella, 2005) but none of them has proved very successful until now.

A natural way to transmit IP multicast over a large geographical area is satellite broadcasting (Tommasi et al., 2010)(Tommasi & C.Melle, 2011).

Among the many applications made possible by the multicast model are those for which security is a critical requirement. Without going too far, the very same IPTV application, when run to pursue economic goals, needs a method to allow only paying customers to access transmitted contents. However many other situations where security is a crucial factor can be imagined (especially in the fields of control and signaling).

According to a recommendation from International Standards Organization (ISO) (*ISO 7498-2*, 1989), while designing a secure system the following criteria are to be considered: confidentiality, integrity, authentication, non-repudiation and access control. To meet such criteria in an IP multicast environment, a Multicast Security (MSEC) Workgroup (*MSEC*, 2011) has been formed within the IETF, with the aim of standardizing protocols for securing group communication over the Internet. Obviously enough, a fundamental topic in the workgroup's activities is the standardization of a group key management architecture. The present paragraph will make use of many of the results coming from the group's efforts and documented so far.

2.1 The multicast group security architecture

The description of the security architecture for IP multicast group communications involves a number of aspects. To reduce the complexity of the presentation, the proposed protocols are grouped in three functional areas, each addressing an aspect of the solution. RFC 3740 (Hardjono & Weis, 2004) outlines the Reference Framework formulated by the IETF Workgroup and identifies such areas (see Fig. 1):

1. *Multicast data handling*. This area includes all the operations on the multicast data performed by the sender and the receiver. Such handling implements:

   - *Encryption*. To support access control and confidentiality, data are encrypted by the use of the group key.

   - *Source authentication and data integrity*. Source identity must be guaranteed by suitable algorithms. Steps are also to be taken to secure the integrity of the received contents.

   - *Group authentication*. This is a minor requirement (guaranteeing the data come from within a group does not necessarily indicate their integrity). However such authentication is very easily achieved and prevents DOS (Denial of Service) attacks.

2. *Group Key Management*. This is the area where secure key distribution and the refresh operations are dealt with.

3. *Multicast Security Policies*. According to (Hardjono & Weis, 2004) Multicast Security Policies represent "the security mechanisms for the group communication" and "the rules for the governance of the secure group".

The Framework also identifies the main elements of a multicast security architecture both in a centralized and in a distributed solution. A central role is played by the "Group Controller" and by the "Key Server". Such entities are usually merged in a single server (GCKS) which is responsible for the "Group Key Management" functional area. Senders and receivers (called GM, Group Members) do interact both with GCKS and with the "Policy server", which is in charge of the "Multicast Security Policies" area.

In order to increase the scalability of the architecture, a distributed approach (see Fig. 1), based on a number of cooperating GCKS, can be opted for. In such case mutual authentication must be guaranteed among GCKS. In a distributed system all receivers will comply with the same security policies and receive the same keys.

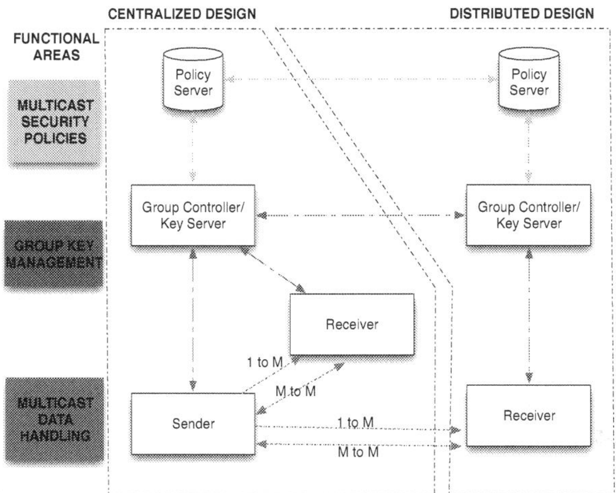

Fig. 1. Multicast Group Security Architecture (from (Hardjono et al., 2001))

2.2 Group Key Management Architecture

The Group Key Management Architecture (Hardjono & Weis, 2004) defines the Group Security Association (GSA) and the main features of the registration and the rekey protocols.

2.2.1 Group Security Association (GSA)

In a protocol designed to manage security on an end-to-end connection, such as IPSEC (Kent & Atkinson, 1998), a Security Association (SA) is a set of shared attributes used by the two ends to secure the connection. Such attributes consist of cryptographic keys, algorithm, identifiers and everything else needed to conduct the communication.

The complexity of a multicast environment imposes the need for more than one key to secure a session. In this context the notion of Group Security Association (GSA) (see Fig. 2) is introduced (Hardjono & Weis, 2004) (Hardjono et al., 2001), which stands for a group of SAs related to the session. SAs in a GSA belong to three different categories:

- REG SA (Registration SA) is used to set up a full-duplex unicast communication channel between GCKS and a GM. GMs start the registration phase by obtaining all needed information directly from GCKS. REG SA is used to protect the other SAs and cannot be set apart from them. It is important to note that no special communication protocol is strictly required here or, for that matter, no communication protocol at all, since a REG SA can even be set up in advance by using a smart card.

- REKEY SA is a multicast security association and it is used to create/renew an SA or to revoke access permission to a GM. It is started by the GCKS with no need of feedback from GMs sharing the same REKEY SA. Contrary to REG SA, it is not always present in GSA. In fact, the lifetime of a group may happen to be so short to make it useless.

- DATA SA (Data Security SA). As for the previous one, no negotiation is needed. It is created by the GCKS to protect the traffic of data flowing from the senders to receivers.

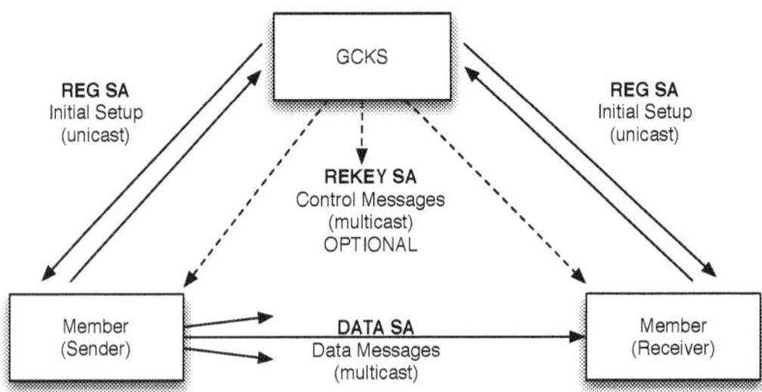

Fig. 2. Group Security Association (GSA) Structure (from (Hardjono et al., 2001))

By using the registration protocol each GM get the authorization and the authentication needed to access a group, to comply with its policies and to obtain its keys. There are two types of keys: Key Encryption Keys, KEK, needed to send keys in a secure way, and Traffic Encryption Keys, TEK, used to encrypt actual traffic. Also a Traffic-Protection Key (TPK) is used, which combines a TEK and a traffic integrity key. KEKs are relevant in a REKEY SA and TEKs/TPKs are relevant in a DATA SA.

2.2.2 Registration protocol

An entity desiring to become a GM will have to use a registration protocol on an unicast connection with the GCKS. The protocol involves mutual authentication between GCKS and the intended GM. When the authentication phase succeeds the GCKS supplies the joining member:

- with all the information needed to start a DATA SA (that is in the case the group security policy requires such a step right at registration and not, as the case may be, as a part of the rekey protocol);

- with all the information needed to start a REKEY SA (provided the group security policy requires a rekey protocol).

Obviously enough, the purpose of the registration protocol is to allow a secure (i.e. authenticated and confidential) transfer of the relevant information between the GCKS and the GM over a SA. Such an SA is called Registration SA. An analogous protocol is dedicated to the purpose of removing the REG SA (in case the GM has not chosen to do it itself).

The design of the registration protocol allows for a good level of flexibility and provides with the ability to support different scenarios. Any secure-channel protocol can be used to deliver the registration messages (e.g. IPsec or TLS). In fact this is what is done with tunneled GSAKMP (Harney et al., 2003). GDOI (see par. 2.4.2) uses IKE Phase 1 to get a secure channel to download REKEY and/or DATA SAs. Authenticated Diffie-Hellman exchanges of the type of IKE Phase 1 are used by protocols like MIKEY(see par. 2.4.3) and GSAKMP(see par. 2.4.1), although they are adapted to increase operations' efficiency.

If for some reason a GM loses the synch with the GSA, it might have to start over a registration with the GCKS. However, there are cases where a simpler method to return in synch may be available:

- the GM can open a plain TCP connection to GCKS and get the recent rekey messages. To open a TCP port to accept such requests might be seen as a dangerous exposition to DOS attacks. In fact, malicious re-synch requests could be an even more serious problem;

- the GCKS could publish the rekey messages on a public (e.g. web) site for the GM to download them from it.

It is desirable that the GCKS provides all three re-synching methods (i.e. new registration, TCP connection, public download).

2.2.3 Rekey protocols

In case of KEK/TPK expiration or group membership changes, the GCKS may update the REKEY SA. A REKEY SA is used to protect rekey messages.
The rekey protocol should possess the following properties:

- rekey information should reach GMs without excessive delays;

- the protocol must specify a way for the GM to contact the GCKS and proceed to a re-synch in case of keys expiration and lack of updates;

- the protocol must avoid implosion problems (see par. 2.2.4) and guarantee reliability in the delivery of rekey information.

The overall scalability of the group key management architecture relies heavily on the performances of the rekey protocol. Therefore scalability must be considered a prerequisite when designing a protocol intended to satisfy the above properties. Rekey protocol should use a scalable Group Key Management Algorithm (GKMA) to send the minimum possible number of keys in a rekey message. LKH (see par. 2.3), OFT (Balenson et al., 2000), Subset difference based schemes (Lotspiech et al., 2001) are examples of GKMA.
A rekey protocol has the following objectives:

- the synchronization of a GSA;

- privacy, authentication (symmetric or asymmetric), replay protection, DOS protection;

- efficient rekeying after changes in group membership or in case of keys (KEKs) expiration;

- allowing GMs to recovery synchronization with GSA;

- a reliable transport of rekey messages;

- good performances in throughput and latency;

- compatibility with multicast and multi-unicast.

A few major issues the design of the protocol must take into account are:

- messages format;

- reliable transport;

- feedback implosion;

- out-of-synch GSA recovery;

- the use of GKMA in rekey messages;

- GKMA interoperability.

2.2.4 Reliable transport of rekey messages

The reliable transport of rekey messages is a crucial point in the design of the protocol.

The content of rekey messages is typically made of KEKs, TPKs, REKEY SA and DATA SAs. Beyond confidentiality and authentication, the protocol must support protection against replay and DOS attacks. GCKS can send the messages to GMs by multicast or multi-unicast.

Confidentiality of rekey messages is obtained by encryption with the Group KEK. If a GKMA is used, the encryption of each part of the rekey message will be performed according to the GKMA specifications, by the pertinent KEKs.

For a GM to receive all intended data it is essential the GCKS is able to keep the SAs (DATA SA and REKEY SA) of such GM in synch. Therefore the reliability of the rekeying mechanism is a fundamental requirement. It can be achieved either by some procedure inherent to the algorithm or by choosing a reliable transport for the rekey messages.

The following solutions have been proposed:

- transmission of multiple copies of the rekey message. It must be recalled that a rekey message may span many IP packets;

- transport by an existing reliable multicast protocol or infrastructure;

- the use of Forward Error Correction (FEC) techniques (together with a feedback carried by NACKs) (Yang et al., 2001).

There is an ample choice of reliable multicast protocols that could be used in our context. While, as of this writing, none of them has started the standard track, a consensus has been reached within IETF on two protocols (Adamson et al., 2009) which are therefore likely to start the track not far from now.

Anyway, no particular reliable multicast protocol has been recommended by the IETF MSEC WG (*MSEC*, 2011) to guarantee reliability in group rekeying. In fact, the choice of the protocol could be subject to special application needs and to the operational environment. Nothing prevents, in the future, the standard use of a particular protocol for the needs of each class of applications.

A major problem arising when using a reliable multicast messaging protocol is implosion. Reliable multicast protocols often make use of ACKs or NACKs to get a feedback about the success of a particular transmission and to start a retransmission in case of failure. Any kind of condition leading to massive packet losses at the receivers can result in the transmission of NACKs from GMs to GCKS. The problem gets soon unmanageable with a large number of GMs. It is referred to as "feedback implosion".

Implosion has been one of the main areas of interest in the topic of reliable multicasting. Some of the solutions proposed to suppress or aggregate the feedback may be well suited in the context of group key management. To reduce the feedback, traffic members may be forced to wait for a random time before sending a negative feedback. During such a wait GMs may receive the needed updates and therefore avoid sending the feedback.

Feedback aggregation is another path followed by some reliable multicast protocols. In this specific domain, however, the concept has drawbacks related to authentication issues. The idea of local recovery (that is establishing local recovery servers to offload the main server) has the same type of problems since GMs should establish SAs with local servers. On the other hand, any subordinate GCKS or even any GM with adequate privileges may act as a local repair server and resend rekey messages.

The main purpose of a GKMA is to make rekeying scalable. Trying to manage a large group without an effective GKMA is plainly unfeasible.

The following points must be kept in mind when selecting a GKMA:

- *Protection against collusion.* GMs and non-members should not be able to join their knowledge in order to discover keys they are not allowed to know (according to GKMA keys' distribution rules).

- *Forward access control.* The GKMA must make sure a GM which has formally left the group is no longer able to re-join it.

- *Backward access control.* The GKMA must make sure when a GM joins the group it cannot decrypt past data.

In order to scale without difficulties GKMAs make generally use of a logical tree structure to organize keys. Obviously there are many ways to manage key trees and to identify a node within a key tree. Within each GKMA packet or at least during the initialization of a REKEY SA the following information has to be provided:

- GKMA name (e.g., LKH, OFT, Subset Difference);

- GKMA version number (implementation specific). Version may imply several things such as the degree of a key tree, proprietary enhancements, and qualify another field such as a key ID;

- number of keys or largest ID;

- version-specific data;

- per-key information:

  - key ID;
  - key lifetime (creation/expiration data);
  - encrypted key;
  - encryption key's ID (optional).

2.3 Logical key hierarchy: a Group Key Management Algorithm

To multicast in a secure an efficient way to a large group of users, a single TEK is generally used for encrypting traffic data. A crucial problem is represented by the users leaving or joining the group. To illustrate it we will refer to figure 3. In theory, for each single change in the users' base, the TEK (K_I in the figure) should be changed. Although grouping changes occurred in a given time interval and updating TEK once for all of them might alleviate the problem, the fact remains that a naive approach would mean transmitting the new TEK to each of the GMs encrypted with the unique key each GM has from its very inception (the KEK, Key Encryption Key, which the GCKS knows for all GMs, from K_A to K_H in figure 3). In other words, the GCKS should send to the group as many copy of the encrypted TEK as the number of GMs. That is an enormous traffic for large groups and, even worse, a constantly repeating one, considering the physiology of the "churn rate". The classical approach to attenuate the problems is that of Logical Key Hierarchy (Wallner et al., 1999) (Wong et al., 1998). Improvements of the LKH (Setia et al., 2000) (Rodeh et al., 1999) (Molva & Pannetrat, 1999) (Zhu et al., 2003) have been proposed with the main purpose of improving scalability of security group associations, in particular during the rekeying phase.

In the tree of figure 4 the circles are keys and the squares are users. The tree has been represented as a balanced binary tree for convenience although there is no particular restriction on its structure. The "leaf" keys are the keys each node has been assigned before

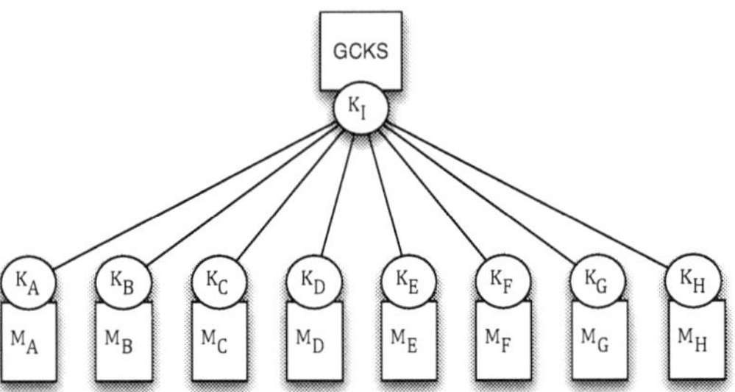

Fig. 3. N Root/Leaf Pairwise Keys

joining the group (e.g. by a smart card). The GCKS must know them beforehand for all nodes. K_O is called root key. By grouping the users in small groups and recursively grouping small groups in larger groups with a number of levels of grouping adapted to the expected total population of nodes, a significant reduction in rekeying traffic can be achieved. Let's see how the basic idea is a GM is supposed to know all the keys on the tree path from itself to the root. For example GM M_{16} must know, beyond its own KEK K_{16}, Key K_H, Key K_L, Key K_N, Key K_O (which is the TEK).

All the intermediate (auxiliary) keys (from K_A to K_N) do not need to be associated with any physical device.

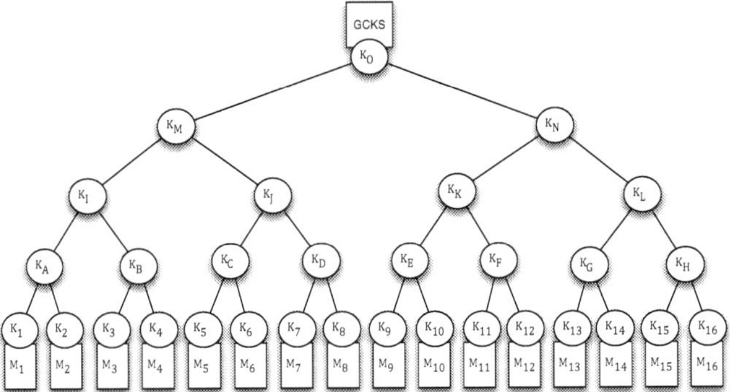

Fig. 4. LKH tree

2.3.1 Join operations

Suppose the user M_3 wishes to join the group. First it will be assigned, in one of many possible ways, its KEK K_3, known only to itself and to GCKS. Next, with some reasonable criterion, it will be associated to a subgroup (the subgroup with KEK K_B in the figure 4). At this point all the KEK from itself to the root (K_B, K_I, K_M, K_O) will have to be changed. They will be the new keys: K_B', K_I', K_M', K_O'. The new keys will have to be known from all the leaf nodes under them in the tree. To have K_B' known to all nodes under it (M_3 and M_4) GCKS will encrypt it with K_3 and K_4. To have K_I' known to all nodes under it (from M_1 to M_4) GCKS will encrypt it with K_B' and K_A. To have K_M' known to all nodes under it (from M_1 to M_8) GCKS will encrypt it with K_I' and K_J. Finally to have K_O' (the TEK) known to all nodes GCKS will encrypt it with K_M' and K_N. The total number of encrypted keys in the rekey message will be $d * log_d(n)$ where n is the number of GMs and d is the degree of the key tree. By this scalable method all the GMs will be able to decrypt the new TEK using the auxiliary keys.

2.3.2 Leave operations

A typical situation is that of a GM leaving the group (e.g. a paying customer of a service willing to unsubscribe from it). The management of the "leave" operation is very similar to that of the "join" one. All the keys previously known to the leaving member will have to be changed in the same way as above.

For a fully populated tree of degree d and height h (where $h = log_d(n)$), the number of keys retransmitted when a member leaves the group is $d * h - 1$ and $d * h$ when a node joins the group (Wong et al., 1998); this compares favorably with the cost of n for a flat system.

2.4 Group key management protocols

A number of group key management protocol have been proposed. Within the multicast security workgroup there are three protocols related to group key management already on the standard track:

- Group Security Association Key Management Protocol (GSAKMP) (Harney et al., 2006). It is intended to be the generic key management protocol and defines methods for policy distribution, policy enforcement, key distribution, and key management.

- Group Domain of Interpretation protocol (GDOI) (Baugher et al., 2003). It uses the ISAKMP phase 1 negotiation as the authentication protocol and sets by it a secure connection between a receiver and the GCKS system. Phase 2 messages are defined within the protocol.

- Multimedia Inter KEYing (MIKEY) (Arkko et al., 2004). It is designed with real-time applications in mind.

2.4.1 Group Security Association Key Management Protocol

The following roles are specified in GSAKMP (Harney et al., 2006):

- Group owner (GO), it is in charge of the policies creation;

- Group Member (GM), it is the end-user (sender or receiver) of all security related procedures;

- Group Controller / Key Server (GCKS), it is responsible for the authentication of GMs, the enforcement of policies, the distribution and management of keys;

- S-GCKS, A GM which can act locally as GCKS when the functions of GCKS are distributed.

Operations of GSAKMP are described for three different scenarios: in the default one a single GM is the sender; in another one (support to which is mandatory) all GMs are potential senders. Support to the third scenario (only a few among the GMs are senders) is left as an option.

In order to enhance scalability, distributed operations are allowed through the set up of local GCKS (S-GCKS). An S-GCKS can provide a better management to its neighboring GMs (e.g. in corporate networks).

GSAKMP operates under the assumption there is at least one PKI (Public Key Infrastructure) for the group to trust. GSAKMP relies on such PKI while creating and verifying security policy rules. The public key of the GO must be known in advance to all GMs.

Upon creation of a new multicast group, the GO starts the process with the creation of a Policy Token (PT) describing the rules for access control and authorizations for that group. The token is signed by the GO. The token contains:

- identification for the PT and group;

- access control rules dictating who can have access to the group keys;

- authorization rules stating who can be a SGCKS;

- mechanisms for handling security, e.g. Security Protocol, Key Creation Method, Key encryption algorithm, Signature, etc.

After a PT is created and signed, it is sent by the GO to a potential GCKS. The latter verifies the signature and, based on the rules specified in the PT, decides whether it can act as a GCKS for the new group. If it can, then the new group is established and all GMs have to register with the GCKS (see Fig. 5). Upon receiving each registration request, the GCKS verifies the signature of the requesting GM and checks whether it is authorized to join the group. If the checks succeeds, the GM receives a "Key download" message. On its part a GM has to verify the GCKS has the authority to manage the group. Eventually, by using the information in the message, a GM can set up both REKEY and DATA SAs. If the GM has no need to send data to the group and it is planning to act as a receiver only, it will have no need to send a "Request to join" message and the "Key download" message is simply sent to the GM after its registration.

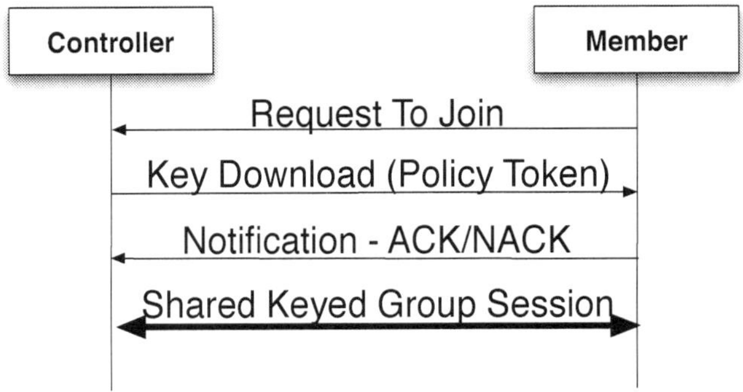

Fig. 5. GM registration in GSAKMP (from (Harney et al., 2006))

A rekeying is required whenever a GM joins or leaves the group and such operation will involve the GO. The latter is informed about node changes and reacts by creating a new PT. PTs must be pushed to the GCKS and the S-GCKS. Upon receiving a PT the GCKS nodes have to check whether the changes involve their own GMs. With no changes, the PT will be distributed according to the LKH by the use of the group key. If some of their nodes has changed than each client must receive the new PT and the only way to do it safely is to encrypt it according to the chosen GKMA and to send everything to every client.

2.4.2 Group Domain of Interpretation protocol

With reference to the ISAKMP (Maughan et al., 1998) terminology, GDOI (Baugher et al., 2003) specifies a domain of interpretation for group key management. While the ISAKMP specification is no longer current (being obsoleted by IKEv2 (Kaufman, 2005)), part of its framework is still used to detail the GDOI specifications.

The setup of secure connections is the result of a two-phases procedure in ISAKMP (and in GDOI). In our terms, the first phase allows to establish a secure unicast connection between the clients and the GCKS. Phase 2 is dedicated to rekeying and the creation of DATA SAs.

Identities of the involved entities are known (together with authorizations) to the GCKS from phase 1 but they can be integrated with certificates provided by the GO in phase 2.

Keys can be transmitted with two formats: GROUPKEY-PULL and GROUPKEY-PUSH. The first one is used by a GM in a client-server fashion to ask for TEKs, KEKs or KEKs arrays (with LKH) according to its needs.

On the other hand, GROUPKEY-PUSH is used by the GCKS when it needs to force the update of the REKEY SA or of the DATA SA.

2.4.3 Multimedia Inter KEYing protocol

The IETF WG has shown a definite interest in the protection of real-time traffic. In particular, the key exchange for SRTP (Secure Real-Time Transport Protocol) has been considered. The MIKEY protocol's design is the result of such focus. It is of use both in one-to-one and in one-to-many exchanges.

The MIKEY (Arkko et al., 2004) protocol specifies key management functionalities. It simplifies the architecture by allowing the sender to incorporate the functions of the GCKS. The Group Control part of the operations, the user's authentication, is performed throughout the course of the initial key exchange by signed messages. The protocol's emphasis on real-time data is represented by its efforts to provide a lower latency, its consideration for the usage over heterogeneous networks and for small groups' interactive exchanges.

The distribution of TEKs is based on the use of either shared keys (distributed in advance) or public keys encryption. With such methods the Traffic Encryption Key Generation Key (TGK) is a shared information between all hosts participating the session. Diffie-Hellman is used for one-to-one connections instead. In this case each client connects to the source (or to the separate GCKS node) and the TGK is different for each GM - GCKS pair.

To avoid the problems associated with the advance distribution of the shared keys, the use of certificates signed by a trusted CA can be preferred. Procedures for rekeying are not defined in MIKEY (the protocol is supposed to be run each time the rekeying is needed). MBMS (Multimedia Broadcast / Multicast Service) (*3GPP*, 2006) is an extension to the protocol designed to allow multicast rekeying in certain environments.

3. Reliable multicast

The topic of the reliable transport of multicast traffic has been already anticipated in par. 2.2.4, especially with reference to the classic problem of feedback implosion. Here we'll present three candidate protocols for the reliable multicast transport of encryption keys.

While protocols based on FEC do generally perform better (Setia et al., 2002) we wish to draw the attention to the fact that in a satellite environment, where the noise tends to be bursty and often a return channel is missing, a protocol simply transmitting multiple copies of the rekey messages might offer a viable alternative.

The three protocols we wish to present are Pragmatic General Multicast (PGM) (see par. 3.1), NACK-Oriented Reliable Multicast (NORM) (see par. 3.2) and our SRDP-Sign (see par. 3.3)

3.1 Pragmatic General Multicast (PGM)

"Pragmatic General Multicast (PGM) is a reliable transport protocol for applications that require ordered or unordered, duplicate-free, multicast data delivery from multiple sources to multiple receivers" (Speakman et al., 2001).

The protocol, developed by a large team of researchers, has the RFC status of "Experimental" as of this writing. Its design puts emphasis on simplicity and does not support much more than the essential capabilities for this class of protocols. Its main concern is the reduction of the repair traffic (driven from NACK implosion or caused by the useless feeding of redundancy to receivers not needing it).

For better operation PGM needs support from the routers crossed by the multicast traffic. That is each router should run PGM-aware software (or firmware) extensions (or, put in different terms, be a PGM NE or PGM-capable Network Elements). At any rate, the protocol can also work, although less efficiently, when some or all of the routers are unaware of it.

PGM runs over the standard IP multicast. As customary with that protocol, GMs can join and leave the group without notifying the source. The only guarantee for a GM is that, once joined the group, it will receive the data with no errors. Any GM can become an independent sender for the group it belongs to and its identification is given by a Transport Session Identifier that no one else can share. PGM is flexible enough to support many different types of applications "as disparate as stock and news updates, data conferencing, low-delay real-time video transfer, and bulk data transfer". Other supplementary options include Designated Local Repairer (DLR) support, fragmentation, late joining, and Forward Error Correction (FEC).

The protocol gets its feedback about the transmission results in the form of NACKs. The potential danger of a NACK implosion is reduced by NACK suppression and NACK aggregation in PGM NE routers (see below).

PGM define the following type of packets:

- ODATA, the Original copy of the transmitted DATA;

- NAK, a Negative AcKnowledgement issued when the receiver realizes a packet is missing in the sequence it received;

- NCF, NAK Confirmation;

- RDATA, a Retransmission of the original DATA;

- SPM, Source Path Message.

3.1.1 PGM transmit window

Mimicking the strategies followed by unicast reliable protocols, PGM keeps a sliding window within which to transmit data. The absence of data allowing to accurately shift the left side of the window leads to the use of a few expedients (based on fixed time waits, on a given period without received NAKs etc.).

3.1.2 PGM tree

To forward data to the intended recipients, PGM builds its own distribution tree (PGM tree) which is identical to the distribution tree natively built by the routers supporting the

IP multicast protocol when all such routers are PGM NE. More generally, PGM builds the distribution tree (the "overlay network") over the original IP multicast tree by having the sender transmitting Source Path Messages (SPM) to the group at regular intervals during the data transfer.

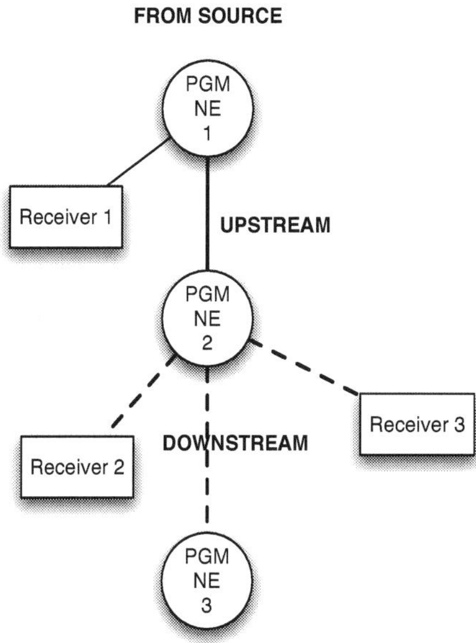

Fig. 6. PGM upstream/downstream attributes for router $PGMNE2$

SPMs are modified at each crossed PGM NE (see Fig. 6). When it reaches a PGM NE an SPM packet contains the address of the PGM NE it comes from. Before forwarding the packet, a PGM NE substitutes its own address to that address so that every PGM NE will always know the address of its upstream closest PGM NE (Gemmell et al., 2003).

When ODATA packets start to flow and a host detects a missing packet, after a random backoff time it sends a NAK to the upstream PGM NE it knows because of the above procedure. On its turn, the latter:

- sends back a multicast NCF packet by the interface that received the NAK;

- forwards to the PGM NE upstream the same NAK packet it received and receives an NCF from it.

The process continues upstream until the source or a DLR is reached. When the NAK reaches the source or a DLR, these may re-send the lost packet downstream to the multicast group, either in the original form or by some FEC encoding.

3.1.3 Local repair: DLR
If no DLR were present, all the repair packets had to be re-sent by the source. The presence of DLRs helps to reduce the outgoing traffic at the source and to limit it to the multicast tree

portion downstream the DLR. It also helps to speed the repair procedure. DLR can announce their presence so that PGM NEs can direct NAKs to them rather than to the source.

3.1.4 PGM with non-PGM-aware routers

PGM can operate even when all routers are not PGM-aware. Of course, with no PGM NE, many of the features of the protocol are lost. For example, each NAK packet will be multicast in the usual way, without the suppression of duplicated NAKs. It will be also impossible to perform efficient repairs, since RDATA packets will be transmitted again and again, no matter how many GM have requested the same packet. The protocol performances will however improve as the number of PGM NE increases.

3.1.5 Congestion control

Congestion control in PGM is performed by limiting the transmission rate at the source. Such limitation is based on the feedback received both from receivers and from PGM NEs. The feedback is given by appending special "report" fields at the end of a NAK packet. The reports communicate the "load" measured by receivers, in the form of packet loss rates, or by PGM NEs, in the form of packet drop rates.

The feedback can be of three different types:

- worst link load as measured by the PGM NEs;

- worst end-to-end path load as measured by the PGM NEs;

- worst end-to-end path load as reported by receivers.

Although congestion control is mandatory, there is no specification of how this data should be used to adjust the sending rate and the choice is left to the implementation (Gemmell et al., 2003).

An extension of the protocol aimed at congestion control has been proposed with PGM CC (Rizzo, 2000). PGMCC is described as "single rate" in that "all receivers gets the same rate and the source adapts to the slowest receiver" and "TCP-friendly" in that the sender tries not to transmit faster than the rate allowed by TCP specifications with the slowest receiver. The protocol adopts a window-based, TCP-like control loop.

3.2 Negative-ACKnowledgment (NACK) Oriented Reliable Multicast (NORM)

According to (Adamson et al., 2009) The Negative-ACKnowledgment (NACK) Oriented Reliable Multicast (NORM) protocol "can provide reliable transport of data from one or more senders to a group of receivers over an IP multicast network". Efficiency, scalability, support for heterogeneous IP networks and for bulk transfers are said to be the goals for the protocol's design. Another interesting target for the protocol is to provide "support for distributed multicast session participation with minimal coordination among senders and receivers". Starting with (Adamson et al., 2009), NORM is on the IETF standard track. In (Adamson et al., 2009) message types and protocol operation are explained in detail. (Adamson et al., 2008) discusses goals and challenges for reliable multicast protocols in general, defines building blocks to address these goals and gives a rationale for the development of NORM.

End-to-end reliable transport of application data is based on the transmission of NACKs from the receivers to initiate repair transmissions from the senders. Variability in network conditions is taken care of by using adaptive timers for the protocol operations. The protocol is designed to offer its transport services to higher levels in a number of ways in order to satisfy the needs of different applications.

NORM uses FEC in various ways. It can use it both in the encoding of the original stream and in the repair traffic sent to the group in response to NACKs from the receivers (proactive/reactive FEC). In general, the more FEC redundancy is put in the original stream the less NACKs will be received.

Most of the potential limitations of the scalability of the protocol come from the negative feedback generated from receivers. NORM uses a probabilistic suppression of the feedback based on exponentially distributed random backoff timers. To avoid disturbing the operations of concurrent transport protocols (e.g. TCP) a congestion control scheme is specified, although alternative choices are left to the implementers.

3.2.1 NORM building blocks

NORM is conceptually divided in three main blocks:

- NORM Sender Transmission, which takes care of data transmissions and reception of feedback (NACK) messages;

- NORM Repair Process, which processes the feedback information and tells the first block what to retransmit;

- NORM Receiver Join Policies, relates to policies and procedures involving receivers admission to the data distribution. While receiver joins are generally unconstrained, a sender might wish to limitate the number of potential NACK senders in various ways.

Other functions (congestion control, error correction etc.) are delegated to further modules.

3.2.2 NORM operations

Messages in NORM are basically divided in sender messages and receiver messages: NORM_CMD, NORM_INFO, and NORM_DATA message types are generated by senders of data content NORM_NACK and NORM_ACK messages generated by receivers within a session.

The NORM_DATA messages are used by senders to transmit application data and FEC encoded repair packets while NORM_NACK messages are generated by receivers to selectively request the retransmission of missing content. NORM_CMD messages are used for various management and probing tasks while NORM_ACK is the acknowledgement message for such commands.

As it is customary in this class of protocols, the receivers schedule random backoff timeouts before sending a NORM_NACK message, which could be repeated if the hoped-for repair has not come. The sender doesn't react to single NACK messages but rather tries to aggregate a number of them to decide how much to "rewind" its transmission. When it deems the rewind to be sufficient, it proceeds to the actual retransmission.

3.2.3 Congestion control

Congestion control for NORM is described in (Adamson et al., 2009). It is an adaptation of the TCP-Friendly Multicast Congestion Control (TFMCC) described in (Widmer & Handley, 2006). It is essentialy based on a rate-control approach rather than on the control of the transmission window. The protocol specification leave, freedom to opt for a window-based approach like that of PGMCC.

3.3 Satellite Reliable Distribution Protocol for Signaling (SDRP-Sign)

The Satellite Reliable Distribution Protocol (SRDP) protocols (Tommasi et al., 2006) (Tommasi et al., 2003) are reliable transport protocols designed with special attention to the use in satellite applications. SRDP-Sign can be seen as an extension of the original SRDP protocol. The two protocols use the same UDP port and implement two different types of transports: SRDP-Bulk and SRDP-Sign. The first one is FEC-based and it is used for bulk data transfers. The second one is of the multi-send type (Tommasi et al., 2008) and has been originally designed for signaling. Despite the original design focus of SRDP-Sign has been the use with short messages (e.g. in multicast control applications) or more generally, with signaling, its relative immunity to burst errors makes it interesting in the context we are examining (Tommasi et al., 2009). One more reason of interest for the protocol is its capability to transmit information to users who can receive information from a satellite but do not possess a return channel. However reliability of transmissions cannot be assured with this subset of users. For all other users, SRDP-Sign is capable of accepting a return feedback both via satellite and terrestrially. The SRDP-Sign protocol is also optimized for a high number of users.

3.3.1 SRDP-Sign: Requirements and architecture

The requirements of the SRDP-Sign protocol are:

- high degree of scalability;

- fast delivery of messages;

- high resistance to burst errors;

- high probability of complete delivery of transmitted data for all users;

- guarantee of complete delivery of transmitted data for users with a return channel;

- limited use of control messaging between sender and receivers.

The objective of each session of the SRDP-Sign is to transmit messages M to R users. Reliability is ensured via transmission of N multiple copies of the messages (Setia et al., 2002). The SRDP-Sign protocol manages the transmission of a single message (SRDP-sign session). The protocol can transmit multiple simultaneous sessions, that is the transmission of the copies of two different sessions to be interlaced. Bundling more messages within a packet is not permitted (see Fig. 7). This preserves the simplicity of packet management.

3.3.2 SRDP-Sign operations

SRDP-Sign ensures reliability of the transmission of a message M, replicating it in N packets. P_i is the i-th reply in the N packets sequence. The delivery of a message is organized in two phases. During the Winding phase, the sender is restricted to replicate the message and there is no control. During the Unwinding phase, the sender makes an estimate of the number of receivers who did not receive the packet during the Winding Phase through a scheme of suppression of the number of receivers.

SRDP-Sign messages can be of the following types:

- DATA, a data packet;

- ABORT, sent to interrupt an ongoing session;

- STAT_REQ, a request to the receivers of a feedback about the correct reception of the message;

- STAT_REP, the answer to a STAT_REQ.

During the Winding Phase the sender transmits N copies (DATA) of the message M. In case of a correct reception of a packet, a receiver ignores all other packets of that message. The replicated packets are transmitted with exponential times that reduces the effects of the potential burst errors (Tommasi et al., 2003).

During the Unwinding Phase the sender multicasts a STAT_REQ to check whether the R receivers have received the message during the previous phase. This request is processed by the receivers through an algorithm of probabilistic suppression of the NACKs. This behavior has a high level of scalability (Nonnenmacher & Biersack, 1998). If a receiver sees the request and has not received the message, then the probabilistic suppression comes into play and if it results in an authorization to proceed, the receiver sends a STAT_REP to the source. A session finishes when the last STAT_REQ message in the sequence (see below) gets no answer. On the other hand, as soon as a STAT_REP message is received by the source, it stops the sending of STAT_REQ messages and proceeds to a new Winding Phase.

The ABORT message, when needed, is also repeated in a fixed way (exactly ten times at regular intervals).

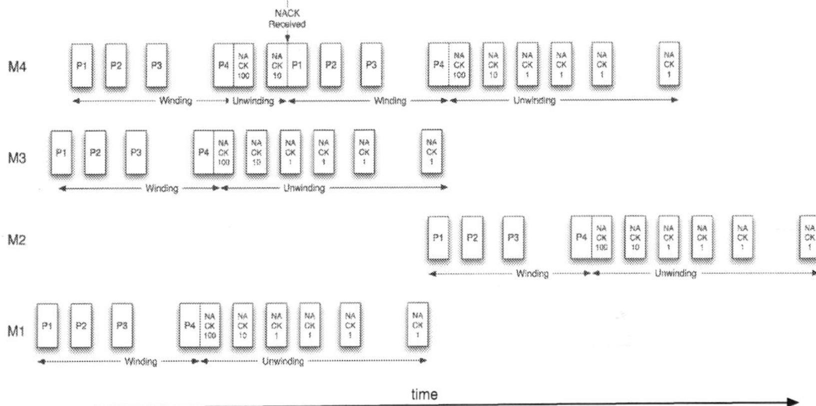

Fig. 7. Transmission of multiple messages (M1 and M2 cannot be interlaced)

3.3.3 Scalable Feedback Suppression (SFS)

Ideally, after transmitting a packet, the sender would like to receive boolean information (yes/no) related to the correct reception from all the receivers. In order to prevent NACK implosion, the Scalable Feedback Suppression (SFS) algorithm causes the random selection of a subset of all the receivers. Such subset is allowed to transmit a negative feedback to the sender. Obviously, only the receivers with a return channel can participate to the selection. The algorithm begins with the selection of an high number H (which represents a very rough and imprecise estimate of the number of potential receivers). The first STAT_REQ transmission is executed and the value probability P_H of 10^r (where $r = -\lceil log_{10}H \rceil$) is included in the message. A receiver that did not receive the original message is authorized to reply only if it generates a pseudorandom value (between 0 and 1) and such value is lower than P_H. If the sender receives even a single NACK, it will abort the Unwinding phase and will re-initiate the Winding Phase (see Fig. 7). If, on the contrary, it does not receive any NACK, the sender iterates the STAT_REQ transmission putting in the message a value of P_H of 10^{r+1}. If the sender does not receive any NACK it increases the transmitted P_H value until it reaches 1. At this point it determines the message has been correctly received by everyone.

4. Performance evaluation

We set up an experimental network to test the performances of the PGM, NORM and SRDP-Sign protocols in different scenarios. Given the scope of the present chapter only a sample of the tests results are reported. The complete results will be the object of a forthcoming publication.

For PGM we selected OpenPGM, an open source implementation available at (*OpenPGM*, 2011). OpenPGM it is not yet a final release. The source code of a NORM implementation is available at (*NORM*, 2007).

The network topology we employed in our test is characterized by an (hybrid) asymmetric connectivity where a single sender is connected directly to the satellite uplink (1Mbit/s) and a small multicast group of receivers has a unicast terrestrial return path to the sender. In this topology, receivers have no access to the satellite uplink but, as it is usually the case, they can receive from the downlink either through a satellite receiver connected to their LAN or by an on-board card (see Fig. 8). The round trip time is about 600ms. We also considered a scenario in which there is no return path to the sender and therefore no kind of feedback is sent by the receivers.

We evaluated the performances of the protocols for various packet loss percentages at the receivers caused by the satellite link. The test is conducted in an homogeneous network with all receivers experiencing the same percentage of independent losses. The packet loss is emulated using Dummynet (Carbone & Rizzo, 2010).

Protocol	Parameter and value	Meaning
NORM	blocksize=64	Number of source packets per FEC coding block.
NORM	parity=32	Number of FEC parity packets.
NORM	auto=32	Number of proactively parity packets.
NORM	unicastNacks	NACK sent in unicast.
OpenPGM	Transmission Group size = 64	Number of source packets per FEC coding block.
OpenPGM	Proactively parity = 32	Number of proactively parity packets.
SRDP-Sign	N=3	number of replies for each message.

Table 1. Configuration Parameters

Protocols configuration parameters used for the present selection of the results are shown in table 1. To put the three protocols on a par, no PGM-aware router has been employed.

We calculated the Average Key Delivery Ratio (AKDR) and the data overhead to evaluate the performances of the above reliable multicast protocols. AKDR is the ratio {number of keys received}/{number of keys transmitted} averaged over all multicast group members. Data overhead is the ratio {total amount of data transmitted}/{net amount of keys data transmitted}.

Fig. 9a shows the results of the tests when a return channel is available, that is the receivers are able to send a feedback to the sender. Fig. 9b shows the results with no return channel available. Despite its simplicity and limited efficiency, it is interesting to note the fairly good

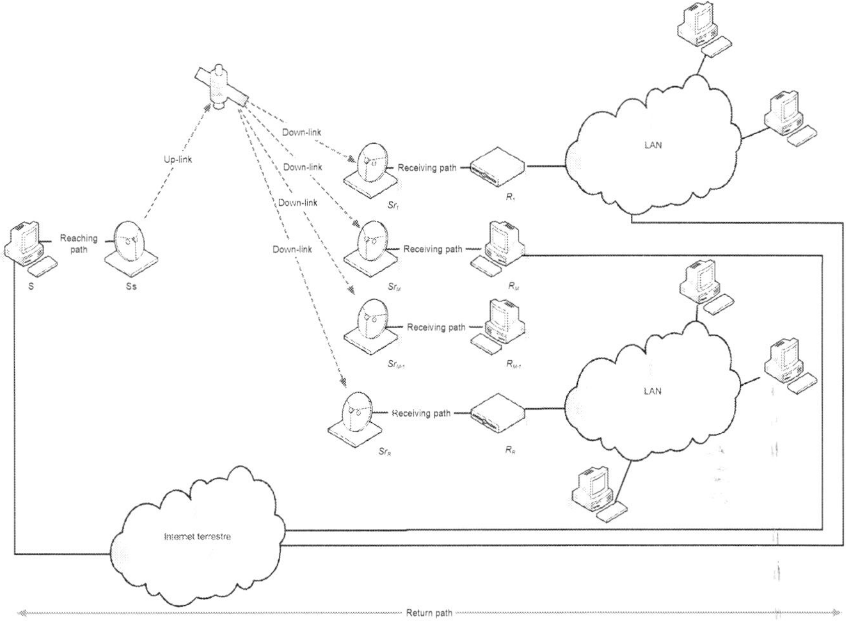

Fig. 8. Test network topology

performances of SRDP-Sign with high packet losses. Increasing redundancy to compensate for high error rates, generally tends to favour the efficiency of FEC based protocols as compared to that of the replica-based ones. However, as our preliminary results suggest, bursty environments (like the satellite ones) tend to level the comparison.

Fig.10 shows a somewhat expected outcome: since SRDP-Sign sends a fixed number of replicas in the winding phase, no matter how much noisy the transmission is, its overhead is by far the largest at low levels of packet losses. On the other hand, when the level of losses increases, also PGM and NORM are forced to retransmit packets, ending up in reaching approximately the same amount of overhead as SRDP-Sign.

5. Conclusions

This chapter introduced the framework and the protocols IETF specified for a multicast security architecture. Three different protocols for key exchange (registration and rekeying), have been presented: GSAKMP, MIKEY, and GDOI. They were developed with different settings in mind, since a single protocol was not believed to be able to support all the typical scenarios in multicast security. LKH is used to allow the rekeying phases to efficiently scale over a large number of users. If the keys are sent via multicast, which is common for large groups and unavoidable with satellites, a reliable multicast transport is required. Three protocols offering such service have been considered: PGM, NORM and SRDP-Sign. The first two of them have been debated within the IETF MSEC WG. The third one was originally conceived for the utilization in multicast signaling (i.e. the reliable delivery of short control

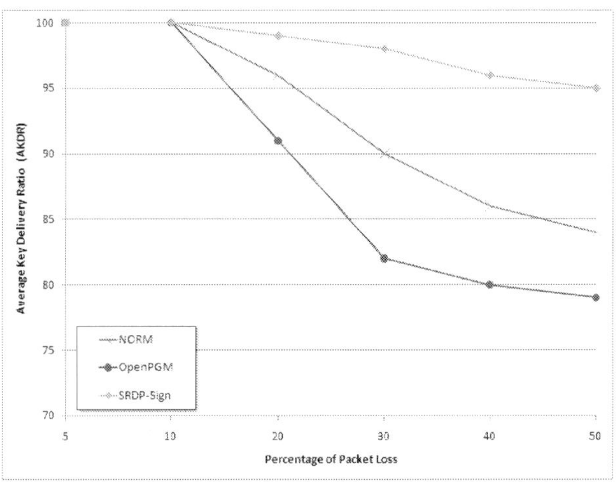

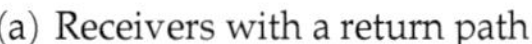

(a) Receivers with a return path

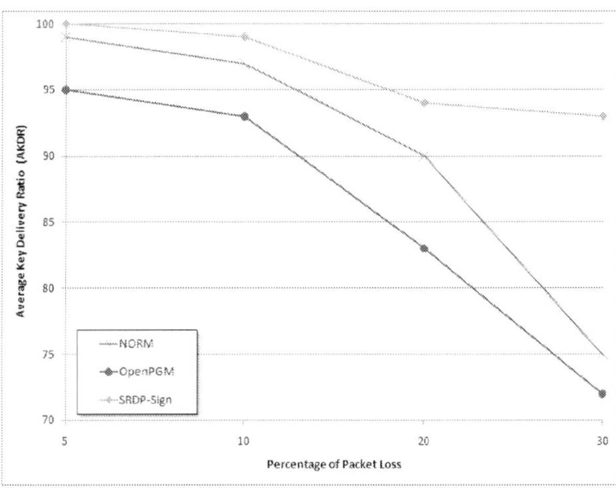

(b) Receivers without return path

Fig. 9. Average Key Delivery Ratio

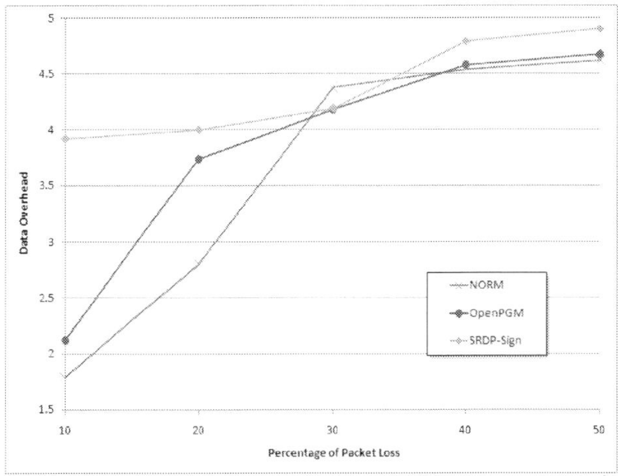

Fig. 10. Data Overhead (receivers with a return path)

messages). However its promising behavior in a satellite environment has prompted to test it in the present context. The preliminary results suggest that, while PGP and NORM do generally perform better, high levels of packet losses (which are typical of the bursty disruption of satellites transmissions) tend to put the simpler approach of SRDP-Sign in a more favorable position.

6. References

3GPP (2006). Security of Multimedia Broadcast / Multicast Service (MBMS). Technical specification TS 33.246.

Adamson, B., Bormann, C., Handley, M. & Macker, J. (2008). Multicast Negative-Acknowledgment (NACK) Building Blocks, RFC 5401. Obsoletes RFC3941.

Adamson, B., Bormann, C., Handley, M. & Macker, J. (2009). NACK-Oriented Reliable Multicast (NORM) Transport Protocol, RFC 5740. Obsoletes RFC3940.

Arkko, J., Carrara, E., Lindholm, F., Naslund, M. & Norrman, K. (2004). MIKEY: Multimedia Internet KEYing, RFC 3830. Updated by RFC 4738.

Arslan, M. G. & Alagöz, F. (2006). Security issues and performance study of key management techniques over satellite links, *Proceedings of CAMAD*.

Balenson, D., McGrew, D. & Sherman, A. (2000). Key Management for Large Dynamic Groups: One-Way Function Trees and Amortized Initialization, Internet Draft (work in progress), draft-irtf-smug-groupkeymgmt-oft-00.

Baugher, M., Canetti, R., Dondeti, L. & Lindholm, F. (2005). Multicast Security (MSEC) Group Key Management Architecture, RFC 4046.

Baugher, M., Weis, B., Hardjono, T. & Harney, H. (2003). The Group Domain of Interpretation, RFC 3547.

Carbone, M. & Rizzo, L. (2010). Dummynet revisited, *Computer Communication Review* pp. 12–20.

Cruickshank, H., Evans, B., Mertzanis, I., Leitold, H. & Posch, R. (1998). Securing multimedia services over satellite atm networks, *International Journal of Satellite Communications* pp. 169–208.

Deering, S. (1989). Host extensions for IP multicasting, RFC 1112. Obsoletes RFC 988, RFC 1054, Updated by RFC 2236.

Deering, S. (1991). Multicast Routing in a Datagram Internetwork, Ph.D. Thesis.

Diot, C., Levine, B., Lyles, B., Kassem, H. & Balensiefen, D. (2000). Deployment issues for the ip multicast service and architecture, *IEEE Network magazine special issue on Multicasting* pp. 78–88.

Dondeti, L. R., Mukherjee, S. & Samal, A. (1999). Survey and comparison of secure group communication protocols, *Technical report*, University of Nebraska-Lincoln.

Eriksson, H. (1994). Mbone: the multicast backbone, *Communications of The ACM* pp. 54–60.

Eskicioglu, A. M. (2003). Multimedia security in group communications: Recent progress in management, authentication, and watermarking, *ACM Multimedia Systems Journal* pp. 239–248.

Gemmell, J., Montgomery, T., Speakman, T., Bhaskar, N. & Crowcroft, J. (2003). The pgm reliable multicast protocol, *Technical report*, IEEE Network.

Hardjono, T., Baugher, M. & Harney, H. (2001). Group security association (gsa) management in ip multicast, *Proceedings of SEC*.

Hardjono, T. & Weis, B. (2004). The Multicast Group Security Architecture, RFC 3740.

Harkins, D. & Carrel, D. (1998). The Internet Key Exchange (IKE), RFC 2409. Obsoleted by RFC 4306, updated by RFC 4109.

Harney, H., Colegrove, A., Harder, E., Meth, U., & Fleischer, R. (2003). Tunneled Group Secure Association Key Management Protocol, Internet Draft (work in progress), draft-ietf-msec-tgsakmp-00.

Harney, H., Meth, U., Colegrove, A. & Gross, G. (2006). GSAKMP: Group Secure Association Key Management Protocol, RFC 4535.

Howarth, M. P., Iyengar, S., Sun, Z. & Cruickshank, H. (2004). Dynamics of key management in secure satellite multicast., *IEEE Journal on Selected Areas in Communications* pp. 308–319.

ISO 7498-2 (1989). Information processing systems, Open Systems Interconnection Basic Reference Model, Part 2: Security Architecture, Internation Organization for Standardization.

Jokela, P. (2006). Key management in ip multicast.
    URL: *http://www.tcs.hut.fi/Studies/T-79.7001/2006AUT/seminar-papers/Jokela-paper-final.pdf*

Kaufman, C. (2005). Internet Key Exchange (IKEv2) Protocol, RFC 4306. Obsoletes RFC2407, RFC2408, RFC2409, Obsoleted by RFC5996, Updated by RFC5282.

Kent, S. & Atkinson, R. (1998). Security Architecture for the Internet Protocol, RFC 2401. Obsoletes RFC1825, Obsoleted by RFC4301, Updated by RFC3168.

Lotspiech, J., Naor, M. & Naor, D. (2001). Subset-Difference Based Key Management for Secure Multicast, Internet Draft (work in progress), draft-irtf-smug-subsetdifference-00.

Mah, F. (2004). Group key management in multicast security.
    URL: *www.tml.tkk.fi/Publications/C/18/mah.pdf*

Maughan, D., Schertler, M., Schneider, M. & Turner, J. (1998). Internet Security Association and Key Management Protocol (ISAKMP), RFC 2408. Obsoleted by RFC 4306.

MBONED (2011). IETF MBONED Working Group.
    URL: *http://datatracker.ietf.org/wg/mboned/charter/*

Molva, R. & Pannetrat, A. (1999). Scalable multicast security in dynamic groups., *Proceeding of the 6th ACM Conference on Computer and Communications Security*.

MSEC (2011). IETF Multicast Security Charter (MSec).
    URL: *http://datatracker.ietf.org/wg/msec/charter/*

Nonnenmacher, J. & Biersack, E. (1998). Optimal multicast feedback, *Proceedings of INFOCOM*.

NORM (2007). NORM implementation Web Site.
    URL: *http://downloads.pf.itd.nrl.navy.mil/norm/*

Noubir, G. & Allmen, L. V. (1999). Security issues in internet protocols over satellite links, *Proceedings of IEEE Vehicular Technology Conference*.

OpenPGM (2011). OpenPGM implementation Web Site.
    URL: *http://code.google.com/p/openpgm/*

Orman, H. (1998). The OAKLEY Key Determination Protocol, RFC 2412.

Rafaeli, S. & Hutchison, D. (2003). A survey of key management for secure group communication, *ACM Computing Surveys* pp. 309–329.

Rizzo, L. (1997). Effective erasure codes for reliable computer communication protocols, *SIGCOMM Comput. Commun. Rev.* pp. 24–36.

Rizzo, L. (2000). pgmcc: a tcp-friendly single-rate multicast congestion control scheme, *Proceedings of ACM SIGCOMM*.

Rodeh, O., Birman, K. & Dolev, D. (1999). Optimized group rekey for group communication systems, *Proceedings of ISOC Network and Distributed Systems Security Symposium*.

Sardella, A. (2005). Video Distribution in a Hybrid Multicast-Unicast World, Juniper networks.
    URL: *http://www.juniper.net/solutions/literature/white_papers/200107.pdf*

Semeria, C. & Maufe, T. (1997). Introduction to IP Multicast Routing.
    URL: *http://www4.ncsu.edu/~rhee/export/papers/multi1.pdf*

Setia, S., Koussih, S., Jajodia, S. & Harder, E. (2000). Kronos: A scalable group re-keying approach for secure multicast, *Proceedings of IEEE Symposium on Security and Privacy*.

Setia, S., Zhu, S. & Jajodia, S. (2002). A comparative performance analysis of reliable group rekey transport protocols for secure multicast, *Proceedings of Performance Evaluation, special issue on the Proceedings of Performance 2002*.

Speakman, T., Crowcroft, J., Gemmell, J., Farinacci, D., Lin, S., Leshchiner, D., Luby, M., Montgomery, T., Rizzo, L., Tweedly, A., Bhaskar, N., Edmonstone, R., Sumanasekera, R. & Vicisano, L. (2001). PGM Reliable Transport Protocol Specification, RFC 3208.

Tommasi, F. & C.Melle (2011). Large-scale terrestrial relaying of satellite broadcasted real-time multimedia streams, *International Journal of Computer Networks & Communications (IJCNC)* .

Tommasi, F., Molendini, S. & Scialpi, E. (2008). Srdp-sign: a reliable multicast protocol for signaling., *Proceedings of NOMS'08*.

Tommasi, F., Molendini, S. & Scialpi, E. (2009). Reliable key distribution for secure multicast by srdp-sign, *Proceedings of AFIN*.

Tommasi, F., Molendini, S., Scialpi, E. & C.Melle (2010). Charms: Cooperative hybrid architecture for relaying multicast satellite streams to sites without a satellite receiver, *Proceedings of IEEE WCNIS*.

Tommasi, F., Molendini, S. & Tricco, A. (2003). Design of the satellite reliable distribution protocol (srdp), *Proceedings of IEEE Globecom*.

Tommasi, F., Molendini, S. & Tricco, A. (2006). The satellite reliable distribution protocol (srdp), *JCOMSS - Journal of Communications Software and Systems* pp. 152–160.

Wallner, D., Harder, E. & Agee, R. (1999). Key Management for Multicast: Issues and Architectures, RFC 2627.

Widmer, J. & Handley, M. (2006). TCP-Friendly Multicast Congestion Control (TFMCC): Protocol Specification, RFC 4654.

Wong, C. K., Gouda, M. & Lam, S. S. (1998). Secure group communications using key graphs, *Proceedings of IEEE/ACM Transactions on Networking*.

Wong, C. & Lam, S. (2000). Keystone: a group key management system, *Proceedings of International Conference in Telecommunications*.

Yang, Y., Li, X., Zhang, X., & Lam, S. (2001). Reliable group rekeying: a performance analysis, *Proceedings of SIGCOMM*.

Zhang, X. B., Lam, S. S., Lee, D.-Y. & Yang, Y. R. (2003). Protocol design for scalable and reliable group rekeying, *Proceedings of IEEE/ACM Transactions on Networking*.

Zhu, S., Setia, S. & Jajodia, S. (2003). Performance optimizations for group key management schemes for secure multicast, *Proceedings of the 23rd International Conference on Distributed Computing Systems*.

Part 5

Sensor Networks

5

Design Issues of an Operational Fire Detection System integrated with Observation Sensors

George Halikias[1], George Leventakis[2], Charalambos Kontoes[3],
Vasilis Tsoulkas[2], Leonidas Dritsas[4] and Athanasios Pantelous[5]
[1]School of Engineering and Mathematical Sciences, City University, London
[2]Center of Security Studies (KEMEA)/Ministry of Citizen Protection, Athens
[3]Institute for Space Applications and Remote Sensing, National Observatory of Athens
[4]Hellenic Air Force Academy, Division of Automatic Control, Athens
[5]University of Liverpool, Department of Mathematical Sciences
[1, 5]UK
[2, 3, 4]Greece

1. Introduction

For the past decades large scale devastating fire events have been occurring in the Mediterranean region. In particular the wider region of Greece has a history of severe fire crisis amounting to devastating damages to property, ecology and losses of civilian lives. High and often abrupt climate variations, hot dry winds, the global warming changing conditions as well as organized criminal activities are the main causes of severe and multiple fire breakouts. These events have resulted in serious crisis situations which often have been developed into natural disasters. Thus, fire events put in danger not only the existing ecological stability of large geographical areas of the country, but also the security of hundreds or even thousands of civilian lives, see also (Marke, 1991) for cable tunnels and his overview of the level of Telecom Australia's operations.

One of the most challenging and serious problems during the evolution of a forest fire is to obtain a realistic and reliable overall common operational view of the situation under development. Combating fires with large fronts is an extremely difficult and dangerous task due to high and abrupt changes of wind direction and intensity, high spatial and temporal variations as well as due to the high variety of forestry and natural vegetation of the environment. In that respect reliable early warning and suppression of fire outbreaks is of paramount importance. Great efforts are made nationwide to achieve early forest fire detection based mainly on human surveillance. These activities are usually organized by the Greek Fire Brigade which is a governmental authority in conjunction with volunteer local private organizations. It is evident that this kind of effort which is based basically on human observation is problematic. Moreover it usually takes place during the summer season between the months of June and September. Our main motivation relies on the fact that to the best of the authors' knowledge at least on a national level there is no operationally sustainable and dedicated sensing network capable of providing reliable early detection and surveillance services to the authorities and public entities. In that respect there are no realistic past data or previous operational experience with similar deployed architectures for fire prevention and monitoring. Another motivation is the recently completed Greek pilot project called: Satellite – based FirE DetectiON Automated System (SFEDONA) which was funded by ESA under ARTES -34 framework (SFEDONA, n.d.; Liolis et al., 2010). For our work we are using as a starting step the basic architectural concept of this project and we proceed further providing an analytical and coherent operational system enriched and extended with Earth Observation components as well as with First Responders critical operational communication sub-systems.

The purpose of this chapter is threefold: At first to introduce commercially available H/W modules that will be useful for fire prevention and monitoring, thus minimizing human intervention actions. Secondly to provide the designer and /or policy maker with some available results from the field of distributed detection theory and how associated methods may be taken into consideration during the initial design phase of the S/W application modules. Thirdly to provide some state of the art technological platforms based on existing and future aerial and space based subsystems that could be properly integrated in the proposed hybrid architecture. In this line and for completeness some important First Responders' open technical communication problems are introduced relating interoperability issues with other existing broadband services. Specific issues related to TETRA communications architectures (Terrestrial Trunked Radio) which are highly critical to the operational capabilities of the search and rescue teams are raised and analyzed. Thus the main contribution of this chapter is the conceptual presentation of an extended operational early warning - monitoring and fire detection system applied but not limited to the Greek situation.

In section 2 some design specifications of the building blocks and subsystems including the satellite communications backbone provided by HellasSat are presented. Further description of the hardware and software components integration is presented in section 3 combined with existing decentralized and sequential detection strategies. A detailed account of the state of the art decentralized approaches in conjunction with some useful fundamental statistical aspects are given adapted to the hybrid model. Different technical limitations imposed during the design and implementation stage are highlighted and presented in section 4. In section 5 some critical operational issues related to communications interoperability between First Responders networks are presented. The integration framework of aerial and space based earth observation remote sensing components with the proposed model is analytically provided in section 6. High-level research directions and guidelines in integrating the proposed architecture with advanced existing and newly developed European space based tools are provided in section 7. It is noted that the introduction and review of the most recent and future technical advancements concerning space based tools in fire and disaster crisis monitoring is presented focusing on the technical efforts of the European Space Agency and the European Community. These efforts are discussed aiming to pinpoint at specific directions for the implementation of an innovative, operational and most importantly sustainable solution. In section 8 the final conclusions of this work are provided.

2. Description of basic system architecture

The proposed model combines terrestrial and space based infrastructures and sensors (SFEDONA, n.d.; Liolis et al., 2010). The terrestrial part is comprised of four general hierarchical levels:

1. The End – Users Fixed Common Operational Center
2. The Remote Fusion/Decision Central Node (**R-F/D-CenN**)
3. The Remote Data Collection Nodes (**R-D/C-N**) and
4. The Set of Environmental Sensors (land based event observers).

The Common Operational Center (CoC) is the public entity or surveillance authority responsible for the coordination and supervision of the overall fire crisis management tasks. The space-based components consist of the Data Handling subsystem (HelasSat) and the Earth Observation infrastructure feeding the Center with all necessary remote sensing earth observations.

The terrestrial platform basically consists of the following IT and Hardware components:

1. Pan-Tilt-Zoom (**PTZ**) cameras combined with the set of environmental sensors and local weather monitoring stations installed at remote and isolated critical areas of interest.
2. Satellite Communication Network based on the DVB-RCS standard along with the installation of the respective satellite terminals for the interconnection of the fixed CoC and the various Remote Fusion/Decision Central Nodes (**R-F/D- CenN**).
3. Earth Observation Imaging Data processing units located at the Operational Center premises.
4. **Wi-Fi** access points (Wi-Fi AP's) for the interconnection of the wireless sensors and PTZ cameras with each one of the Remote Data Collection Nodes (**R –D/C-N**).
5. **Zig-Bee** (IEEE 802.15.4 standard protocol) - to - **WiFi** (IEEE 802.11) gateways providing links between the ZigBee network of the wireless environmental sensor set and the rest of the WiFi network.
6. Independent Power Supply Units (such as small Solar Panels) for the energy powering of the Remote Data Collection Nodes (**R-D/C-N**) and the Remote Fusion/Decision Central Nodes (**R-F/D-CenN**).
7. The above system components combine standard protocols with available Commercial - Of - The Self (COTS) products. A careful selection must be made so that various performance criteria and trade offs are met such as: interoperability, quality attributes, format of the component, necessary physical resources for the functioning of each device component, technical limitations and restrictions, capacity, size, performance specifications, data handling etc.

The IEEE 802.15.4/ZigBee protocol for Wireless Sensor Networks allows fast, scalable and easy network deployment and adoption supporting QoS combined with COTS devices and technologies. It is known that the IEEE 802.15.4 Data Link/ZigBee network layer, allows the implementation of three network topologies - Star, Mesh and Cluster- Tree - while ZigBee defines 3 types of devices: ZigBee Coordinator (**ZC**) -ZigBee Router (**ZR**)- ZigBee End Device (**ZED**), see for more details (Cunha et al., 2007; Da Silva Severino, 2008). It is noted that a key feature of the IEEE 802.15.4 Data Link/ZigBee devices is the classification into two subcategories: The Full Function Devices and the Reduced Function Devices. The later are End Devices implementing only very simple (reduced) applications such as infrared passive sensing (IR passive sensor devices) transmitting very small amounts of information in the sensor network. Thus they are very beneficial in terms of low power consumption since end – devices can be asleep for long periods of time and can wake up only when it is needed for data transmission.

In the sequel various software component applications are proposed to run on the sub-systems such as:

- Application to run at the Common Operational Center premises for continuous monitoring, surveillance and control of the remote geographical regions.
- Application to run at the Operational Center for immediate alerting of the end-users in case of fire breakouts. An integrated powerful Geo-Spatial Information Subsystem (GIS-subsystem) for fire representation and spreading is proposed to support decision making on the part of the end-users, indicating the exact location of the fire events using available vector/raster background maps.
- Intelligent Software application to run locally at the Remote Fusion/Decision Central Nodes for event - observation and fusion of critical heterogeneous data coming from different sensing sources such as wireless PTZ cameras and the Remote Data Collection Nodes. Additionally advanced intelligent software applications will be necessary for decentralized event detection, and fast decision policy making.
- Application to run at the Remote F/D Central Nodes for critical data and fire alarm communication/transmission to the CoC.
- Application to run at the Remote Data Collection Nodes for simple local decision-making and message re-transmission. This type of node due to more relaxed power constraints compared to the set of environmental sensors' stringent power constraints, should be capable of more advanced signal processing/decision capabilities on a local level.

The selection criteria of the software components regarding intelligent algorithms for observation fusion and fast event detection is probably one of the most challenging tasks for this type of distributed networks. Several design and modeling issues related to this problem are addressed and discussed in the sequel. Additionally various performance indexes are introduced for performance evaluation of the detection algorithms. A short review of the current literature results and design efforts of intelligent decentralized detection is provided.

3. Analytical component description

As it is seen in Fig.1 below the basic component blocks are:

1. **The Satellite link**: This link provides a two-way data transmission with high reliability between the Remote Fusion/Decision Central Nodes located at the geographical areas of interest and the CoC which is located at the end user's location (village municipality or city). Both terminals operate in dual mode (receive and transmit) were for fire event detection and alerting the uplink data transmission from the Remote F/D Central Node to the CoC is of primary importance. For the Greek *terrain* and *environment* the baseline scenario involves the communications infrastructure concerning the GEO Ku-band satellite HellasSat2 at 39 deg. E. and its operational network which is based on the **Digital Video Broadcasting-Return Channel via Satellite (DVB-RCS)** standard. HellasSat owns and operates the Hellas Sat-2 geostationary satellite which provides IP and DVB services and thus will establish the backbone satellite communications link. In case of fire events detection and alerting data, messages are transmitted to the end user via the satcom interface. For the purpose of an alert verification or fire in progress situation, a low frame rate video stream can be transmitted to the site of the end user. A relatively low data rate satellite link is required and an assumption of 512/256 Kbps is reasonable.

2. **Remote Sensing:** Remote Sensing coupled with advanced information, telecommunication, and navigation technologies contributing to high-speed geo-spatial data collection more efficiently than ever, and supporting the disaster management organizations to work with higher volumes of up to date information. Fire imaging from remote platforms can be used in emergency response for strategic and tactical operations. Strategic observations are provided mainly by polar orbit satellite systems like NOAA/AVHRR, MODIS, ENVISAT, etc. These observations are in different spatial and spectral resolutions, and give a regional view of fire occurrences with time intervals ranging from some hours to one day. These observations are useful for disaster coordination support but are ineffective for repetitive timely observations due to orbit cycles. On the other hand tactical operations, which need real time observations, are efficiently served by the geostationary orbit satellites like the Meteosat Second Generation, as well as airborne (manned or unmanned) platforms that are able to provide continuous coverage and rapid data accessibility over the entire country and individual fire events respectively. Among the most enduring data flow bottlenecks existing today are the challenges for interoperability during the operations, where space/airborne remote sensors need to work together with in-situ sensor networks and data fusion and processing nodes as it is proposed in our network architecture.

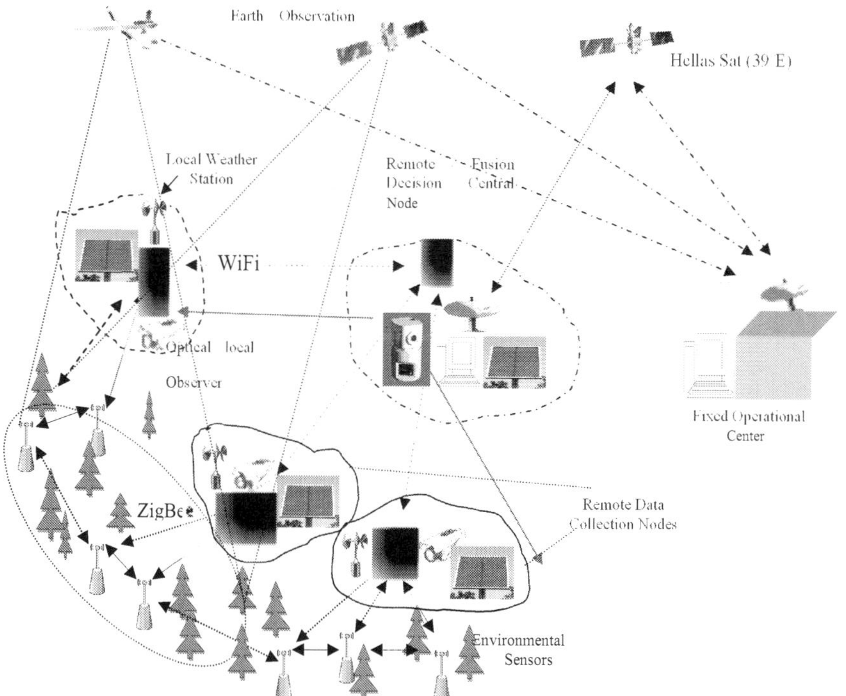

Fig. 1. Early Warning and Fire Detection Physical Model Architecture.

3. **Common Operational Center site**: The fixed CoC is located at the end user's location (municipality, community or village site) and consists of an integrated S/W platform. The platform is responsible to provide common operational and continuous centralized and remote monitoring of the critical areas that are under surveillance and inspection. Moreover in case of fire detection or alarm signaling the CoC immediately is informed with the support of an integrated GIS - application for the indication of the exact location of the fire event with the aid of available digital maps of the region. The S/W application will include fast intelligent computational algorithms for real time estimation of fire front propagation based on inputs streaming from the environmental wireless network (**Remote Fusion / Decision Nodes**). Additionally the end users will be able to remotely control the PTZ cameras installed at the Remote F/D Central Nodes. In that manner continuous monitoring of the critical geographical sectors will be possible through video sequences/frames coming from the on field camera sensors. A satellite terminal also is needed to be installed providing satcom links between the field and the CoC.

4. **Remote Fusion/Decision Central Node (R-F/D-CN)**: The remote fixed fusion/decision central node will be physically located at a safe distance from the critical area of interest such as a forest or an area of high probability for a fire event to occur. It is responsible mainly for data fusion and decision-making as well as for alert distribution. It is noted that the final decision-making process and strategy for event detection is taking place at this Central Node by performing a probabilistic likelihood ratio test. It is based on the received observations and partial local type decision outputs of the Remote Data Collection Nodes. In that respect decentralized detection is of major importance and it is comprised of two main parts, see (Fellouris & Moustakides, 2008; Chamberland & Veeravali, 2007):

   a. The sampling strategy at the remote sensors (Event Observers).
   b. The detection policy at the fusion – decision center which in our case is the Remote Fusion / Decision Central Node.

Policies related to sampling rates basically define the type of sensor data that is transmitted to the Remote Fusion/Decision Node while policies related to detection concern the utilization of the transmitted information by the Fusion/Decision Node such as the final decision of the occurrence or not of an event. Sampling/detection strategies performed at the fusion center is a discipline of ongoing research efforts. The concept of decentralized detection was first introduced by (Tsitsiklis, 1993) and later by (Veeravali et al., 1993). We mention that for the centralized detection case the fusion center has complete access to the continuous time process observations which in our application set up is the fire event spatio-temporal evolution. Additionally one or (usually) several R-F/D-Central Nodes may be installed depending on the geographical region and terrain morphology.

Analytically the following components are required:

- A satellite terminal so that communication between the Remote - F/D - Central Node and the Operational Center is possible.
- A WiFi access point with its integrated controller so that communication between the various heterogeneous data coming from optical cameras, environmental sensors and small local weather monitoring stations is achieved. The link is based on the IEEE 802.11/WiFi family standards for short/medium range communications at the frequency of 2.4 GHz. The data rate can be up to 25 Mbps.

- A panoramic PTZ camera which will act as a redundant fire detection and surveillance device adding additional degrees of freedom to the overall architecture. As soon as the distributed sensors such as the optical cameras, the local weather stations and the environmental sensors generate an alert of a fire event, the end user can remotely point and zoom the PTZ camera to the specific site thus getting a better and fullest picture of the situation. PTZ cameras have the technical ability to pivot on their horizontal and vertical axis (pan/tilt head) allowing the users to cover and survey the areas of interest. Also they have an automatic setting allowing for scanning on a predetermined axis.

Some technical specifications are given depending on the product type and cost (COTS):

  - Horizontal scanning ability of 340 deg. and vertical scanning of 100 deg.
  - Motion sensors.
  - IR sensitivity for scanning at nighttime or areas of low light.
  - Ethernet cable connectivity.
  - Up to 45 frames/sec at a resolution of 640x480.
  - JPEG & MPEG encoding.

- An integrated S/W application so that data fusion and decision policy tasks are possible. The input to the F/D central node includes all available information generated from the sensing hardware such as optical cameras, environmental sensors and the local weather monitoring stations. Moreover this S/W application will be responsible for data and alarm transmission to the Operational Centers' site.

- An independent power supply unit. It is necessary since the Remote –F/D-Central Node will have to be autonomous and as it is mentioned will be located at strategic geographical areas of possible high-risk fire events. In that way independence from the power grid is achieved. The power unit is proposed to be based on relatively small solar panels, including charging battery arrays, inverters and controllers, so that autonomous operation is achieved for several days.

5. **Remote Data Collection Node (R-DC-N):** It is the basic Data Node responsible for collecting the various data sequences transmitted from the local optical cameras and environmental land based sensors and for performing the real-time local sensing of the fire event. There can be one or more data collection nodes depending on the application and geographic location or sector and will be located near or inside the critical areas in fixed positions. Moreover an additional assumption is made that these nodes are capable of re-transmitting an amplified version of its own local partial observation and local decision of the events at the remote central node. Thus the remote data collection nodes act in the network as amplifiers transmitting sequences of finite alphabet messages to the Remote Central Nodes. Furthermore each node consists of the following sub-systems:

- **A WiFi access point with an integrated controller** for the communication between the Remote Data-Collection Node and the Remote F/D Central Node. The Data collection node will be collecting all available data from the sensors: optical camera, local weather monitoring stations, and the environmental sensors and feed them to the Remote F/D Central N. The interface data rate can reach up to 25 Mbps.

- **Wireless optical cameras or "Optical Observers"** responsible to perform real time local detection of fire-smoke-flame parameters. It is proposed that embedded image processing algorithms are included or further developed depending on the morphological terrain. The operation of these "Optical Observers" is mainly based on low-resolution high dynamic range contrast camera providing robust representation of an event or scene under various uncontrolled illumination conditions. Reliable and advanced image/signal processing algorithms can be implemented either of the self (COTS) or by in house development. The range coverage of each camera will be in the order of a few Kilometers. For, each Data Collection Node four "Optical Observers" suffice to cover wide geographical regions (>30 km).

- **A remote weather monitoring local station** attached to the mast of each sensor - optical camera capable of collecting and monitoring local weather information such as: Wind direction and speed, humidity factors, air temperature variations, dryness etc. Such an "instrument" will add extra degrees of freedom and reliability when combined with the optical "Observers" and the environmental field sensors. It will provide valuable information related to the status of a potential fire event or even further to provide critical information for fire front prediction and progress. The proposed weather stations will be able to transmit data to the Remote Fusion/Decision Central Node via the WiFi access point (Wi-Fi AP) of the Data Collection Node.

- **A Zig-Bee (IEEE 802.15.4) to WiFi gateway** which will be attached to each optical camera. It can act as a local coordinator of the ZigBee network of the environmental sensors that are associated with a specific camera. Basically the gateway establishes a bridge between the ZigBee network topology of the environmental sensors installed in the distant area and the Remote Data Collection Node. In that way critical data coming from the installed wireless sensors can further be communicated via the WiFi wireless interface to the Remote F/D Central Node. The interface data rate can be up to 115.2 kbps from the ZigBee side and up to 25 Mbps from the WiFi side.

- **An independent Power Supply subsystem**. By definition and due to the distant location, functionalities and hardware limitations, the Data Collection Node will have to operate autonomously with no human intervention and totally independent of the power grid network. The operation will need to be proper and seamless for long periods of time. In that respect this power subsystem will provide power to the components of the Node such as the WiFi access point, the optical sensors, the remote local weather station and the ZigBee to WiFi gateway. A solar panel based power system is proposed that is similar to the one mentioned for the Remote F/D Central Nodes. However since there will be no satellite terminal at the site of the Data Collection Node the power specifications and constraints can be significantly relaxed.

- **The wireless local environmental sensors** that are distributed in the areas of interest measuring parameters such as humidity, smoke, flame, temperature or soil moisture. These sensors are low cost, low power and small size wireless devices capable of having communication links between them at low data rates via the ZigBee wireless network interface. The density of the distribution (distance between the sensors) strongly depends on the morphological terrain of the critical site of interest. A typical distance could be 1 km in the field areas. Communication of the environmental sensors with the rest of the network is feasible using the Zig-Bee to WiFi gateway which is attached to each optical camera installed at the Data Collection Node. Finally an option for the installed sensors could be the family of Low Power RF transmitters. With respect to the three known Zig-Bee topologies star-mesh-cluster-tree as are shown in Fig. 2., for the application of fire detection the selection should be made between the mesh and cluster/tree topologies since cluster/tree topologies provide higher network flexibility, and are more power efficient using battery resources in a more optimal fashion.

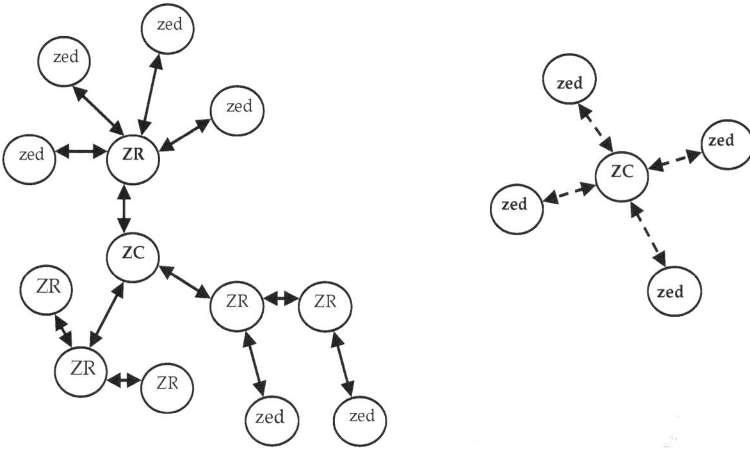

Fig. 2. Zig-Bee typical Cluster – Tree and Star network topologies.

An important performance testing parameter to be accounted for in the specified network application is the "smooth" coexistence of WiFi/ZigBee technologies since they both operate at the same 2.4 GHz band. For example minimization of interference risks is of paramount importance since it has been observed in various applications that ZigBee can experience interferences from Wi-Fi signal traffic transmission (some packet losses due to increased WiFi power levels). Careful field testing investigation is required when employing the wireless network to evaluate and confirm the coexistence limits of both types of RF based technologies.

In the following Fig. 3., a Top Level - schematic geometry of the early warning architecture is presented. The Earth Observation components are excluded for simplicity reasons.

4. Distributed event detection strategy methods

For distributed land based sensor networks related to fire detection and environmental monitoring the event can be characterized as rather infrequent. In that setting surveillance is highly required while reliability and timeliness in decision-making is of paramount importance. Thus decentralized rapid detection based on fusion technology and intelligent algorithms play a key role in the proposed model (Gustaffson, 2008). In particular decentralized detection is an active research discipline imposing serious research problems and design issues, [Bassevile & Nikiforov, 1993; Chamberland & Veeravali, 2006, 2007) and (Tsitsiklis, 1993; Veeravali et al., 1993). In the proposed application low cost flame, smoke, and temperature detectors as well as additional local environmental sensors to be employed are subject to various power limitations. The classical concept of Decentralized Detection introduced by [Sifakis et.al., 2009] considers a configuration where a set of distributed sensors transmit environmental finite-valued messages to a fusion center. Then the center, is responsible for the decision making and alerting while the classical hypothesis testing problem is solved deciding on one of the two hypotheses, that are *"a change has occurred"* or *"a change has not occurred"* see (Gustaffson, 2008).

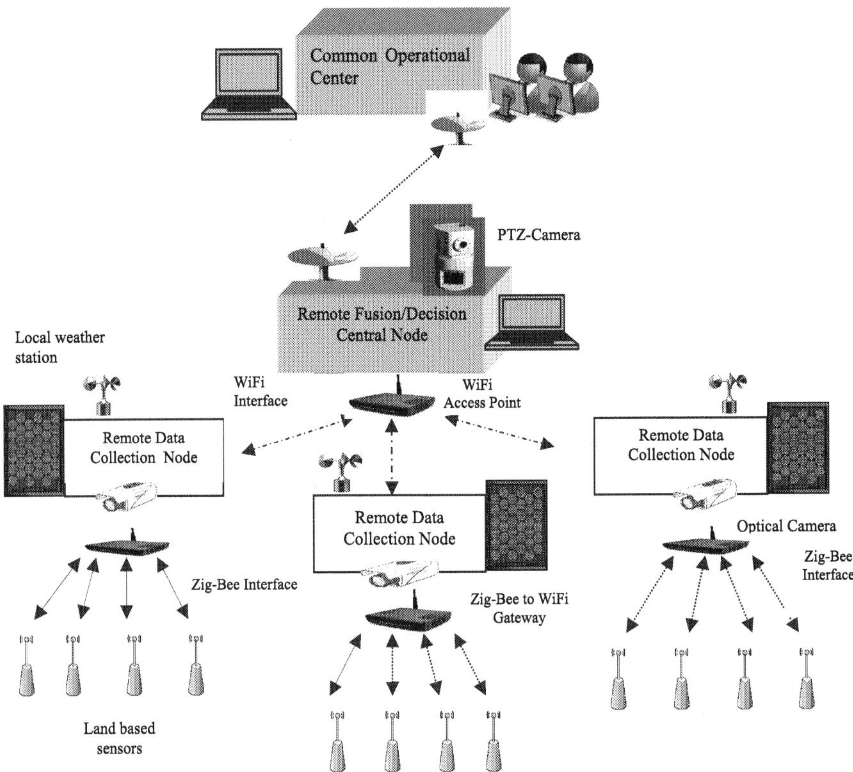

Fig. 3. Block interface schematic diagram of the proposed model without the earth observation components.

In our operational application , the decision on the type of data and alarm sequences to be sent to the Common Operational Center is primarily realized at the Remote Fusion/Decision Central Node (Global Decision Maker) since the Remote Data Collection Node acts more as a concentrator of field data capable of taking some kind of partial local decisions. It is noticed that the proposed model is decentralized in contrast to traditional centralized configurations where each distributed sensor communicates all observation data to the fusion center (most optimal case but with no design constraints). Moreover it is assumed that all data collection nodes can take decisions using identical local decision rules (Chamberland & Veeravali, 2006).

As stated in (Tsitsiklis, 1993), decentralized schemes are definitely worth considering in contexts involving geographically distributed sensors. Also in (Chamberland & Veeravali, 2006), it is explicitly stated that the basic problem of decentralized inference is the determination of what type of information each local sensor should transmit to the fusion center. It is evident that efficient design of a sensor fire detection/surveillance network depends strongly on the interplay between data compression, available spectral bandwidth, sensor density of the network and resource allocation, observation noise, and overall optimal performance of the distributed detection process. Moreover for the decentralized case the collected observations are required to be *quantized* before transmitted to the central fusion node. These *quantized* measurements then belong to *a finite alphabet* . This procedure as it is mentioned previously is the result of a combination of technical specifications such as stringent communication bandwidth and high data compression. For the proposed system each sensor transmits its own partial observation parameter such as smoke or flame, to the Remote F/D Central Node and thus it is sub-optimal when compared to centralized schemes were the central node has direct and full access to all observation sets. Careful and detailed analysis is necessary for the adoption (or in house development) of intelligent algorithms at the remote central fusion decision node.

Moreover realistic assumptions need be taken into account related to the shared medium or the so-called common wireless spectrum. As it is pointed out, in (Imer & Basar, 2007), several performance design challenges need to be combated when designing wireless networks such as limited battery power, possible RF interference from other sources, multipath effects etc. The restriction on batteries life cycle of the low RF power transmitters or the power supply is of major importance and imposes severe limitations on the duration of time each sensor is going to be awake/on and the number of transmission cycles is capable of making. In our case Data Collection Nodes are autonomous and backed up by solar panel power devices. On the other hand the different low cost environmental sensors scattered in the remote areas impose hard power limitations. Issues such as Optimal Measurement Scheduling with Limited Measurements need to be considered when developing the detection algorithms both at the Fusion/Decision Central Node and at the CoC site. In (Imer & Basar, 2007; Fellouris & Moustakides, 2008), the problem of estimating a continuous stochastic process with limited information is considered and different criteria of performance are analyzed for best finite measurement budget.

At this point, we mention design issues imposed by channel fading and attenuation. In a realistic situation the quality of the communication channels between the environmental sensors and the remote data collection and fusion/decision units is affected and degraded by heavy environmental changes, bad weather conditions, heavy noise and disturbances, different SNR's, bad location dependent connections etc. Design parameters related to the channels state and fading level need to be included during the design stage, see for further details (Chamberland & Veeravali, 2006; Imer & Basar, 2007).

Another important twofold issue is the type of observations at the sensors and the sensor location and density. A popular assumption is that these observations (or data) are conditionally independent which might not hold if sensors are to be distributed with close proximity and high density in a specified area. In that scenario sensors will transmit observation data that are strongly correlated. Then the theory of large deviations can be employed to evaluate the performance of the network. In our case as it is previously mentioned the environmental sensors can be employed at least within a distance of a few hundreds of meters apart of each other. It is not well known a priori what distance will produce correlated or uncorrelated observations. This depends on how large the fire front will be or of the fire progress in general. As it is explicitly stated in (Chamberland & Veeravali, 2006) the optimal location of the sensor network before deployment requires careful analysis and optimization and it involves a design tradeoff between the total number of nodes and the available power resources/node of the network.

Furthermore a realistic assumption for the observation data is that they are conditionally independent and identically distributed see (Chamberland & Veeravali, 2006; Gustaffson,

2008; Bassevile & Nikiforov, 1993) for further detailed exposition. Then assuming that there are resource constraints, optimality is assured using identical sensor nodes. Optimality under this type of condition is a positive fact since these networks are robust and easily implementable. In the figure below the *conceptualization of a decentralized detection model* is presented.

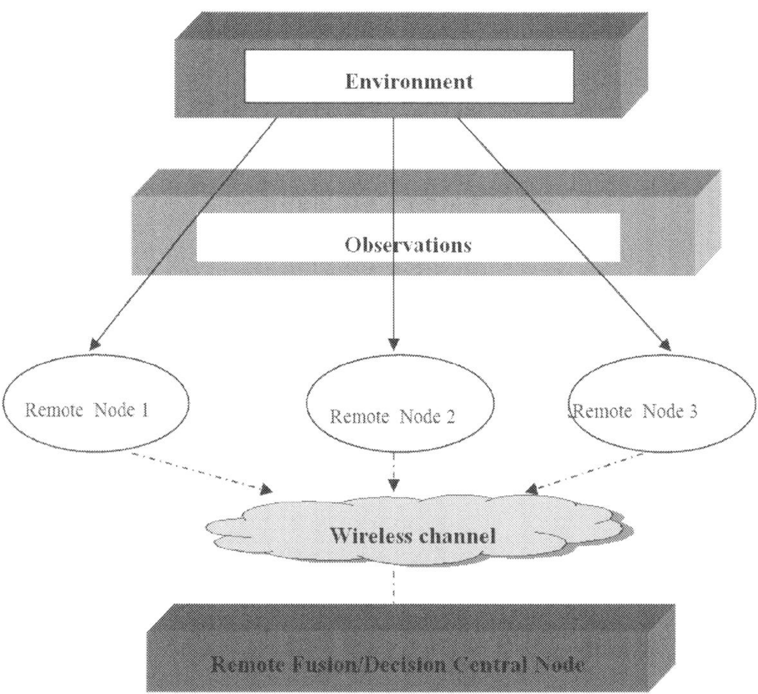

Fig. 5. Conceptual geometry of a decentralized detection model.

It is evident that both the number of transmitted data per node and the number of available nodes is finite as well and the finite alphabet constraint is imposed on the output of each sensor. Then the basic problem that needs to be solved at the Remote Fusion/Decision Central Node is of a statistical inference type.

Another important design issue is that of decentralized sequential detection which for our system is carried out as previously stated at the Central Node. Sequential detection and hypothesis testing strategies involve deep mathematical results and various algorithms have been successfully applied in modern state of the art change detection and alarm systems.

In typical change point detection problems the basic assumption is that there is a sequence of observations of stochastic nature, whose distribution changes at some unknown time instant λ, for $\lambda = 1,2,3,\ldots$. The requirement is to quickly detect the change under false alarm constraints. For the distributed case at hand, as it is shown in Figure 3, measurements are realized at a set of L distributed sensors. The sensor's outputs can be considered in general as multi-channel and at some change-point λ, one channel at each sensor changes distribution. Since sensors transmit quantized versions of their observations to the fusion center, change detection is carried out. At this point it is useful to mention some very basic facts and definitions related to On-line Detection. The subject enjoys intensive ongoing research since wireless and distributed networks are in fact gaining great popularity with an abundance of applications such as the one considered in this work.

Let $(y_k)_{1 \le k \le n}$ be a sequence of random variables with conditional density $f_\theta(y_k / y_{k-1},\ldots,y_1)$ and θ be the conditional density parameter. Before an unknown time of change t_0 the parameter $\theta = \theta_0$ (constant). After the change time, the parameter θ assumes the value θ_1 and the basic detection problem is to detect this change as quickly as possible. Then a stopping rule is needed to be defined which is often integrated in the family of change detection algorithms. Moreover an auxiliary test statistic g_n and a threshold λ is introduced for alarm decision. The typical *stopping rule* has the basic form $t_a = \inf\{n : g_n(y_1,\ldots,y_n) \ge \lambda\}$ with $(g_n)_{n \ge 1}$ being a family of functions of n coordinates and where t_a is the so-called *alarm time* that the change is detected see (Bassevile & Nikiforov, 1993) for an extensive account. More formally the definition of a stopping time is the following:

A random variable (map) $T : \Omega \to \{0,1,2,\ldots,;\infty\}$ is called a stopping time if

$$\{T \le n\} = \{\omega : T(\omega) \le n\} \in \mathbb{F}_n, \quad \forall n \le \infty, \tag{1}$$

or equivalently

$$\{T = n\} = \{\omega : T(\omega) = n\} \in \mathbb{F}_n, \quad \forall n \le \infty \tag{2}$$

Notice that $\{\mathbb{F}_n : n \ge 0\}$ is a filtration, that is an increasing family of sub-sigma algebras of $\mathbb{F}$. Finally five fundamental *performance criteria* are presented which have an intuitive reasoning to evaluate and assess change detection algorithms:
1. Mean time between false alarms,
2. Probability of false detection,
3. Mean delay for detection,
4. Probability of non-detection,
5. Accuracy of the change time and magnitude estimates.

Usually a global performance index concerns the *minimization of the delay for detection for a fixed mean time between false alarms*. For the proposed fire detection set up it is important that careful analysis of available sequential detection algorithms is performed taking into account the above criteria as well as the basic tradeoff between two measures: detection delay and false alarm rate.

A series of statistical tests for continuous time processes (such as the Sequential Probability Ratio Test - SPRT and the Cumulative Sum - CUSUM test) exist which can be combined with state space recursive algorithms such as the Kalman filter or adaptive filtering techniques for change detection and state estimation of the fire evolution (Gustaffson, 2008).

These tests are fully performed at the Remote Fusion/Decision Central Nodes as well as at the Common Operational end user's site of the proposed architecture. It is well beyond the scope of this paper to further analyze this class of algorithms and techniques and how they are integrated and implemented in fire detection software applications. Nevertheless any early fire warning and monitoring system should consider carefully the above design and software component issues, see (ESA, 2008; Tartakovsky & Veeravali, 2004).

Finally it is stressed that in the current literature, assumptions include discrete samples (binary messages) and synchronous communications between the fusion center and the sensor devices. The approaches concerning continuous time processes require additional sampling/ quantization policies. For example fire and flame flickering is time varying and can be modeled as a continuous random process (Markov based modeling approach). In these cases and due to power and transmission constraints the Remote F/D Central Node receives *data in a sequential fashion* and the goal is to quickly detect a change in the process as soon as possible with a low false alarm rate. On the other hand bandwidth limitations require *efficient sampling and quantization strategies* since canonical or regular sampling may no longer be optimum.

5. Integration with First Responders communication systems

It is important in this subsection to take a step further and raise the complex issue concerning First Responders (FRs) needs with respect to communications interoperability extending the scope of the proposed fire detection/surveillance system. This aspect which in our opinion is not usually addressed in various proposed detection/surveillance systems is highly important and operationally critical to any designer who needs to consider a fully realistic high level integrated architecture. In the case of large fire disaster and crisis outbreaks it is highly probable that first responders teams from other European nations and various local emergency response entities will be involved in the crisis monitoring and mitigation efforts. Thus serious interoperability problems of the dedicated heterogeneous communications subsystems will arise due to different communication standards. Indeed at the technological level the variability of available technologies that are used among First Responders networks result in a diversity of characteristics such as signal waveforms, data throughput, latency and reliability, and security (i.e. different cryptographic standards). This situation results in serious compromise of coordination and operational efficiency among FRs even at the monitoring level of the events. Moreover it is well known that at a European and national level different Public Safety authorities have adopted different systems, equipments and often dedicated technology resulting in a multitude of networks which are non-interoperable. Thus interoperability is in fact a critical factor for European Public – Safety and Security teams that deal with an environment that is complex, interconnected and highly interdependent. We only mention dedicated networks such as Professional Mobile Radios and TETRA/TETRAPOL networks. These networks function under different architectures and air interfaces and so internetworking (roaming capability) is extremely difficult. Additionally new technical capabilities are continuously being adapted by FRs such as ad-hoc mesh broadband networks which are able to provide and extend connectivity over the affected areas of interest and to deliver high data throughput which can be higher than 5Mbs. In Figure 4 a simplified schematic is provided of different FRs with the associated isolated networks.

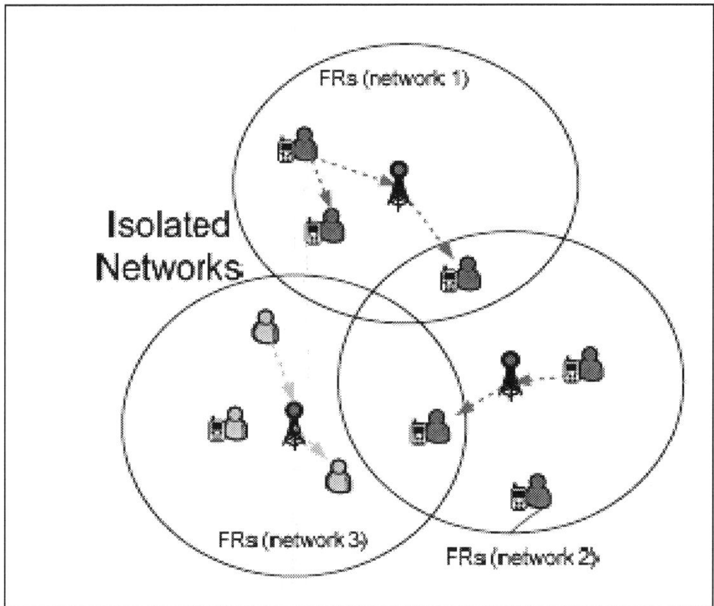

Fig. 4. First Responder isolated communication networks.

TETRA has been transformed from circuit switched to IP packet switched architecture (IP protocols) for more efficient integration with other existing technologies. An open design implementation problem then is to account for short term like solutions that will be able to interconnect most existing communication sub-systems and networks using a possible dedicated node ensuring interoperability of all systems without the need to modify existing equipment such as handset devices and other communication infrastructures. In that manner FRs will be able to continue to use current receiver equipment, communication base stations and other critical infrastructures. Thus a specialized gateway could be a possible unifying and cost effective alternative for technical interoperability between different FRs networks capable of supporting across network - services (**cross-network services**) such as : Voice-calls between TETRA, TetraPol and WiMAX broadband networks, exchange of location based data, exchange of images or seamless transmission of emergency broadcast signals over heterogeneous networks to the specific geographical area of interest or the exchange of a high-priority information across networks. Another issue to be addressed during the design phase is security adopted to critical situations. There are well established techniques and methods (e.g., RSA, DES/3DES, AES encryption) that guarantee security across networking. Nevertheless these type of measures can become a serious problem during a major Fire event since security policies may prohibit communication across different FRs networks. In the same context we mention the existing technical problems related to interoperability even when the same technology is used within a country such as communication between TETRA – TETRA systems.

For the case of Greece TETRA is the dominant technology used by emergency and surveillance authorities. This is also the case for most European countries. In particular

TETRA is replacing legacy-PMR technologies, to become the most common technology to use or it is being considered for future adoption where an emerging associated challenge is the additional spectrum requirements for all TETRA future networks as well as Inter-System-Interface (ISI cross border communications). We briefly mention some basic TETRA key-services such as: Registration, Authentication, Individual Half duplex Call, Priority Call, Preemptive Call (emergency), Broadcast Call, Instant Messaging. Also other early – adopters are already experimenting with the use of broadband technologies such as WiMAX or extension of current PSC coverage. In addition as is the case of the proposed Fire Detection Operational System the exploitation of Satellites for backbone communications infrastructure is especially critical since it provides seamless connectivity between the critical geographical area of interest and the Common Operational Center. This type of space based links is used by the majority of FRs of most European member countries while cellular technology is used as a complementary means. For the hybrid model the S-band satellite services could be used for integration and connectivity so dual use of TETRA/S-band terminals can be exploited providing data rates up to 10Mbs, or a dual mode S-band/L-band terminal providing data rates up to 500Mbs.

In conclusion when designing the architecture of an operational fire detection and monitoring system technical aspects related to the integration with existing FRs communication systems must be addressed and cannot be ignored even at a conceptual level. These include: Interoperability of different networks based on standard protocols (TETRA, TETRAPOL, PMR and WiMAX) or between networks of the same technology (TETRA – TETRA). Interconnection of various full-duplex/semi-duplex networks (such as GSM, ISDN e.t.c.), Air-Interface aspects of each different network technology such as the existing base stations or radio terminals, Network management functions of decentralized networks, connectivity and full integration with satellite systems.

6. Integration of operational observation platforms

In this subsection we propose specific state of the art sub-systems that can be integrated in the proposed model, as they have reached such a maturity level that may rank them between the operational tools in the emergency response. These components mainly constitute more advanced earth observation and space based subsystems and assets such as ESA's Earth Observation program and ESA's and EC Global Monitoring for Environment and Security program, the so-called GMES program, with its component supporting risk management and emergency response (ESA 2008, 2009; NOA, 2007). We mention here space and airborne-based surveillance tools and more specifically early warning and near real time monitoring systems with integrated fire risk and fire mapping modeling capabilities using:

a.   Medium to Low-Resolution Remote Sensing sensors.
b.   High-Very High Spatial Resolution Remote Sensing for detailed mapping and damage assessment, and identification of critical infrastructures prone to fire risk.
c.   Airborne thermal sensing platforms.

Several studies show that despite the low spatial resolution of the order of a few kilometers, the SEVIRI instrument onboard the MSG satellites, offer high potential for real time monitoring and disaster management. According to (Roberts et al., 2004) there is a considerable correlation between the fire radiative energy and the corresponding signals captured by the SEVIRI and MODIS sensors. Due to this (Umamaheshmaran et al., 2007) and (Van den Bergh & Frost, 2005) exploited the high update rate of the MSG/SEVIRI images and showed that the use of image mining methods improves significantly the information extraction from MSG/SEVIRI in view to detect fires and model the fire evolution.

With the occurrence of the disastrous wildfires of summers in 2007 and 2009 in Greece, the Institute for Space Applications and Remote Sensing of the National Observatory of Athens (ISARS/NOA) deployed its MSG/SEVIRI fire monitoring service, in complement to the existing operational emergency response state capabilities, providing support to decision makers during the fire fighting operations. Today the MSG/SEVIRI fire monitoring service of ISARS/NOA is offered on a 5-15 minutes basis supporting the actions of a number of institutional civil protection bodies and fire disaster managers all over Greece. With this service the rapid identification of new fires arises has become possible within an average alert time of 5 – 20 minutes. However, there are limitations relating to the instrument's low spatial resolution and geo-location accuracy; due to its distant geostationary orbit (i.e., 36 000 km) and the renown resolution limitations of thermal sensors, the MSG/SEVIRI has a ground sampling distance of the order of 4 km over Greece, which, theoretically, allows for the detection of wildfires with a minimum detectable size of about 0. - 0.30 ha see (Prins et al., 2001). Nonetheless, the elevated saturation temperature (>335 K) in the SWIR band minimizes the saturation effect allowing for a sub-pixel fire characterization. This means that, due to the important temperature contrast between the hot spots and the background, outbursts sizing much smaller than the nominal resolution of the sensor may also be detectable under certain conditions as it was the case in all deployed fire monitoring operations in Greece. However, if we want to meet the existing early warning and timely fire detection needs, these figures may not comply with standard detection requirements of fires, the later being approximately 2-3 times smaller, namely 0.1 ha see (Rauste at al., 1999). For this, although the MSG/SEVIRI data are, for the time being, the only satellite data that can be used to improve the reliability in fire announcements, because of their low spatial resolution, they cannot be used alone but as a network component, the later integrating a variety of other sensors as proposed in this paper. It is obvious however that a space based monitoring component as the one of ISARS/NOA, may affect significantly the sensor network topology and lead to high simplifications, especially when the network needs to be deployed in large geographic areas with much accentuated topographic relief as in Greece.

Referring to space based monitoring capabilities it should be noted however that much higher spatial resolution representations can be offered from a number of polar orbit satellite systems like SPOT, LANDSAT, IRS, IKONOS, FORMOSAT-2, etc. However, the main difficulty with these systems is the fixed orbit geometry of the satellite platforms, which results in restraints in revisiting capability both in tactical operations, and in surveillance of vulnerable areas prone to high risk. In contrast, aircraft (manned or unmanned) are much more easily maneuverable and may very quickly revisit the critical areas providing rapid response for emergency situations. Airborne TIR sensors are usually FLIR (Forward Looking InfaRed) cameras, capable to detect new hot spots that develop rapidly into wildlands. Besides aircrafts equipped with FLIR sensors can be used for supporting fire-fighters in safety tasks, and for detecting escape routes or security zones, in areas where the human visibility is restricted due to the smoke.

For this purpose ISARS/NOA developed and is capable to deploy on demand an airborne fire sensing service under the name SITHON see (Kontoes et al., 2009a). In reality it makes one component of a larger network of sensors, as the whole SITHON system comprises a wireless network of in-situ optical cameras, coupled with the airborne fire detection platform of NOA/ISARS. This network is linked to an integrated GIS environment in order to facilitate real time image representation of detected fires on detailed background maps, that incorporate qualitative and quantitative information needed to estimate the prone to the risk areas and help the disaster management operations (e.g. fuel matter, road network, morphology, endangered locations, endangered critical infrastructures like fuel stations, flammable materials, industrial areas, etc). Moreover, the platform of SITHON includes a Crisis Operating Centre, which receives information in the form of images and data from the wireless sensor detection systems, displays it on wide screen monitors and analyses it to derive the dynamic picture of fire evolution. The airborne system is designed to ensure automatic fire detection. It is mountable on any airborne platform and can be operated within 15 to 20 minutes after the first fire announcement. Once on the platform, SITHON is supported by a fully automated control system, which manages the frame acquisition, the radiometric image calibration and signal thresholding, as well as the dynamic fire detection and geo-positioning within 50-100 m error using on board GPS and INS technology and with the lack of any operating GPS station on the ground. The minimum fire size detectable by the system can be of 3x3 meters on the ground from 2000m Above Sea Level (ASL). The integration of the NOA/ISARS airborne monitoring component in the proposed network topology as indicated in figure 1, enhances the monitoring capacity of the sensor network and improves the automatic fire detection and terrain surveillance capability in geographically extended areas. In the following Figure 5, we provide the SITHON platform. A 310Q CESSNA two-engine aircraft.

Fig. 5. SITHON / Platform – airborne imaging system. (Reproduced picture from (Kontoes et al., 2009a)).

7. Future research directions

Future research directions could definitely include the integration with ESA's Data Dissemination System DDS, the other polar orbiting systems such as EnviSat and **GMES** Sentinel spacecrafts, the integration of UAV sensors, which can provide real time data transmission to the ground, and the improvement of algorithms and models used for raw data processing, and data fusion and analysis of space, aerial, and terrestrial observations, to obtain higher detection accuracy and timely announcements of fire alarms. Moreover new fire detection algorithms need to be explored and validated accounting for the local specificities, morphological features and land use/land cover conditions of the area they apply. To this end NOA/ISARS has proposed improvements in the algorithmic approaches proposed by EUMETSAT for fire detection using Meteosat Second Generation satellites, and introduced appropriate adaptations over Greece to avoid fire model detection uncertainties and reduce the returned false fire alarms, see (Sifakis et al., 2009).

At this point, it is briefly mentioned that our proposed model could further be extended and integrated with the web based European Forest Fire Information System consisting of two operational sub-modules: The European Forest Fire **Risk** Forecasting **System (EFFRFS)** which is a module for fire risk forecasting information and processing and the European Forest Fire **Damage Assessment System (EFFDAS)**, which is capable of evaluating and assess the damage caused after a fire event using satellite imagery.

Furthermore, two additional elements could be certainly proposed for integration in the proposed architecture for future deployments: Unmanned **A**erial **V**ehicles (**UAV'**s) for surveillance and monitoring tasks especially for large-scale fire events and ESA's new initiative of a Satellite Based Alarm System. The latter case needs further intensive technical efforts (such as the identification of appropriate frequency selection and interoperability aspects) taking advantage of the current GSM/UMTS systems for broadcasting messages to mobile phone users in dedicated geographical regions were the fire event is taking place. UAV sensors capable of carrying IR and video cameras and instrumentation with high-resolution capabilities for dedicated fire and hot spot detection, as the airborne SITHON observing system presented above, it seems very promising for reliable and fire monitoring services see (ESA, 2008; Kontoes et al. 2009a; 2009b; 2009c). More explicitly they can serve concurrently several tasks such as vegetation mapping and forestry, fire fighting and emergency management airborne communication collection and relay, as well as environmental monitoring before and after the fire event. With such systems further localization and confirmation of fire sources in conjunction with the proposed fire detection system, can be achieved therefore minimizing significantly the false alarm rate. We mention that this type of systems and their integration with existing space and terrestrial infrastructures are currently under ESA's research efforts. Indeed co-operative Satellite - UAS missions can deliver unrivalled global area coverage and time-critical, very close range operational capabilities (ESA, 2008; 2009). Even more in the near future the European **Data** **Relay Satellite System (EDRS)** will be a reality and further integration with the above components will be an attractive space based sustainable solution. The EDRS system offers (and will be technically capable in offering) real-time or nearly–real time response times for rapid information updating and Rapid Mapping activities and Surveillance including the "very urgent" imaging data downlink as well as meeting the growing demand for "<1 meter" resolution data availability (ESA, 2008).

Finally we should mention that in the case of Greece, several initiatives namely RISK-EOS, SAFER and LinkER - are run by the National Observatory of Athens – Institute for Space Applications and Remote Sensing, funded by the European Space Agency and European Union within the GMES program framework (Kontoes et al., 2009b; Robertson et al., 2004). These initiatives foresee the provision of additional services that respond but not limited to, wild fire crisis management in the entirety of Greece. In particular the central and basic set of core services provided during the crisis are near real time fire mapping (the so called rapid mapping) at high and very high spatial resolution, as well as continuous monitoring and early warning on a 15 minutes basis using medium to low spatial resolution satellite derived products. These services are offered through dedicated gateways of GMES, making appropriate use of properly developed interfaces linking the local End User community and the corresponding GMES National Focal Point, that is the National Observatory of Athens with the Emergency Response Core Services (ERSC) gateway.. The main aim is to rapidly assess and disseminate information on fire occurrence and combine it with additional in-situ and space/aerial collected data to effectively support early warning, as well as decision-making and coordination of the emergency response actions during fire fighting. The integration of these newly developed operational geo-information services in the framework of GMES, to the proposed architecture is an innovative element providing complementary fire detection and fire mapping information that needs to be considered for future directions, in the implementation of more reliable and integrated fire warning and monitoring architectures. In fact a large-scale deployment of the proposed system in various geographical areas of Greece could be well complemented by the integration of additional fire occurrence and fire spreading evidences through NOA's established monitoring capabilities and GMES/ERCS gateway (Kontoes et al., 2009c).

8. Conclusion

In this chapter the basic model architecture for timely and accurate fire detection and surveillance according to operational user requirements is described. Hardware and software issues as well as satellite, airborne and terrestrial data handling technologies have been described and their integration to the proposed network observing architecture is justified. Some important and mission critical communication issues related to First Responders Network interoperability were also provided. These issues are of high priority when it comes to further integrate and extend the proposed system with the response emergency authorities on a national and international level. Additionally the integration of Earth Observation platforms is commented and their integration was presented. Moreover some important theoretical aspects of decentralized detection strategies were provided.

Time is the most crucial parameter in fire combating and fire containment. The level of efficiency depends on the promptness of the detection system to receive and send in almost real time its alarming signals indicating fire outbreaks and fire locations. The state-of-the-art in most of the deployed fire sensor systems, seem not to take this into account, namely various aspects related to sequential change detection design parameters and optimality issues arising in decentralized detection schemes over wireless communication channels, as proposed in this paper. On the other hand part of the existing literature regarding distributed detection systems is strongly theoretical and involves esoteric and often deep results from the fields of statistical estimation and sequential change detection theory.

This work concludes to an operational and realistic, in terms of efficiency and cost of deployment, initial modeling solution, and ensures that the proposed model is easily expanded to the newly developed and emerging Earth Observation, Telecom, Navigation, Aviation and Advanced Sensor technological advancements, in order to efficiently address the problem of early detection and prompt emergency response in the case of fire disasters. The disaster management community will be soon facing a great technological peak, enabled by the advancements in aviation, sensor and imaging technologies, telemetry, data fusion and processing, and geo-information/value added products use. The authors are currently involved in assisting the integration of these technologies to the daily practice of the disaster management community through on-going research and development in the domain of state-of-the-art integrated application systems.

9. References

[1] Bassevile, M. & Nikiforov, I.V. (1993). *Detection of Abrupt Changes: Theory and Application*, Information and System Sciences Series, ISBN 0-13-126780-9, Englewood Cliffs, N.J.

[2] Chamberland J.F. & Veeravalli V.V. (2006). How Dense Should a Sensor Network Be for Detection with Correlated Observations?. *IEEE Transactions on Information Theory*, Vol. 52, No.11, pp. 5099-5106.

[3] Chamberland, J.F. & Veeravalli, V.V. (2007). Wireless Sensors in Distributed Detection Applications. *IEEE Signal Processing Magazine*, pp. 16-25.

[4] Cunha, L.J.; Alves, Q. & Koubaa, M. (2007). On IEEE 802.15.4/ZigBee to IEEE 802.11 gateway for the ART-WiSe architecture. *IEEE Proc. of Emerging Technologies and Factory Automation, ETFA*, pp. 1388-1391.

[5] Da Silva Severino, R.A.R. (2008). On the Use of IEEE 802.15.4/ZiGBee for Time – Sensitive Wireless Sensor Network Applications. *Report of the Instituto Superior De Engenharia Do Porto*

[6] European Space Agency (2008). Internal documentation on Satellite Systems and Operations for Unmanned Aerial Systems ESA/JCB, *Advanced Research on Telecom Systems* (ARTES).

[7] European Space Agency (TLTP 2008). *Telecommunications Long Term Plan (2009-2013)*, ESA/JCBc 47, rev. 7.

[8] European Space Agency. *Internal documentation ESA/JCB*. (2009). Advanced Research on Telecom Systems (ARTES).

[9] Fellouris, G. & Moustakides, G.V. (2008). Asymptotically optimum tests for decentralized change detection. *Proc. of the International Workshop on Applied Probability, IWAP2008*, Compiegne, France

[10] Gustaffson, F. (2008) *Adaptive Filtering and Change Detection*, John Wiley & Sons, ISBN 0471-49287-6

[11] Imer, O.C. & Basar, T.(2007). *Wireless Sensing with Power Constraints*, Springer-Verlag Berlin Heidelberg: C. Bonivento et al. (Eds): Adv. In Control Theory and Applications, pp. 129-160.

[12] Kontoes, C.C.; Keramitsoglou I.; Sifakis N. & Konstantinidis P. (2009a). SITHON: An Airborne Fire Detection System Compliant with Operational Tactical Requirements, *Sensors*, Vol. 9, pp. 1204-10.

[13] Kontoes C.C.; Poilvé H.; Florsch G.; Keramitsoglou I. & Paralikidis S. (2009b). A Comparative Analysis of a Fixed Thresholding vs. a Classification Tree Approach for Operational Burn Scar Detection and Mapping, *International Journal of Applied Earth Observation and Geoinformation*, Vol. 11 No.5, 2 pp. 99-316.

[14] Kontoes C.C.; Sifakis N. & Keramitsoglou I. (2009c). GMES Burn Scar Mapping kicks into full gear after 2007 wildfires in Greece, Windows on GMES, A BOSS4GMES Publication, No. 3, pp. 58-63.

[15] Liolis, K.P.; Pantazis, S; Gennatos, V; Costicoglou, S; & Andrikopoulos, I. (2010) "An Automated Fire Detection and Alerting Application based on Satellite and Wireless Communications", *Proc. 5th Advanced Satellite Multimedia Systems Conference*

(ASMS2010) & 11th Signal Processing for Space Communications Workshop (SPSC2010), Cagliari, Italy, September 13-15, 2010.

[16] Marke, P. (1991). Cable tunnels - an integrated fire detection/suppression system for rapid extinguishment, *Fire Technology,* pp. 219-233.

[17] National Observatory of Athens NOA. (2007) – Institute for Space Applications and Remote Sensing, *RISK EOS, extension to Greece.*

[18] Prins, E.M.; Schmetz, J.; Flynn, L.P.; Hillger, D.W. & Feltz, J.M. (2001). An Overview of Diurnal Active Fire Monitoring Using a Suite of International Geostationary Satellites. Global and Regional Vegetation Fire Monitoring from Space: Planning a Coordinated International Effort, edited by Ahern F.J., Goldammer J.G., Justice C.O., Hague, The Netherlands.

[19] Rauste, Y.; Sephton, A.J.; Kelhä, V.; Vainio, T.; Heikinheimo, M.; Soini, K.; Frauenberger, O. & San Miguel-Ayanz, J. (1999). Forest Fire Operational study: Requirements and Analysis Report RAR, VERSION 2.3 (AO/1-3468/98/I-DC). *Report to the European Space Agency.*

[20] Roberts, G.; Wooster, M.J. & Perry, G. (2004). Fire Radiative Energy: Ground and Satellite Observations. *Geostationary Fire Monitoring Applications Workshop,* EUMETSAT.

[21] SFEDONA (n.d.) Available from: http://telecom.esa.int/telecom/www/object/index.cfm?fobjectid=29777 ESA project.

[22] Sifakis, N.; Iossifidis, C.; Kontoes, C. & Keramitsoglou, I. (2009). Wildfire detection and monitoring over Greece using MSG-SEVIRI satellite data (submitted).

[23] Tartakovsky, A.G. & Veeravalli, V.V. (2004). *Change -Point Detection in Multichannel and Distributed Systems with Applications, In: Applications of Sequential Methodologies* (N. Mukhopadhyay, S. Datta and S. Chattopadhyay, Eds), Marcel Dekker, Inc. N.Y., pp. 331-363.

[24] Tsitsiklis, J.N. (1993). Decentralized Detection. *Advances in Statistical Signal Processing,* Vol. 2, pp. 297-344.

[25] Umamaheshwaran, R.; Bijker, W. & Stein, A. (2007). Image Mining for Modeling of Forest Fires From Meteosat Images. *IEEE Transactions on Geoscience and Remote Sensing,* Vol. 45 , No. 1, pp. 246-253.

[26] Van den Bergh, F. & Frost, P.E. (2005). A Multitemporal Approach to Fire Detection, Proceedings of the 2nd IEEE International Workshop on the Analysis of Multi-temporal Remote Sensing Images, pp. 156- 160.

[27] Veeravali, V.V.; Basar, T. & Poor, V.H. (1993). Decentralized Sequential Detection with a Fusion Center Performing the Sequential Test. *IEEE Transactions on Information Theory,* Vol. 39, No. 2, pp. 433-442.

Part 6

High Capacity Satellite Communications

6

Passive Microwave Feed Chains for High Capacity Satellite Communications Systems

Giuseppe Addamo, Oscar Antonio Peverini,
Giuseppe Virone and Riccardo Tascone
IEIIT –CNR Torino,
Italy

1. Introduction

The successful implementation of satellite communication systems requires robust wireless channels providing the up-links and down-links for the communication signals. The frequency operative bands employed depend on the particular application. Navigation and mobile satellite systems are typically operated in the L (1-2 GHz) and S (2-4 GHz) bands, whereas remote-sensing applications are mostly offered in C (4-8 GHz) band. In the commercial communication area, due to the increasing demand of high quality services, the operating frequency bands has evolved towards the Ku (12-18 GHz), K (19-21 GHz) and Ka (27-32 GHz) bands. Although communication systems operating in high frequency bands provide more channel capacity, the effect of free-space attenuation and atmospheric absorption can limit the performances of these systems (e.g. signal-to-noise ratio). In this contest, the employment of efficient transmission algorithms and protocols provide meaningful advantages, but the bottle-neck is however represented by the antenna system that has to satisfy very strict requirements. For these reasons each device composing the antenna-feed chain has to be designed in order to guarantee significant electromagnetic performances and, at the same time, high integration levels (Cecchini et al., 2009). Moreover, when high power levels are employed (also of the order of tens of KW), further problems are related to spurious interferences generated by non-linear devices, as microwave amplifiers, and by metallic contacts that behave as a diode junction due to the oxidation of the metals. Additionally, high-power and low-pressure conditions can cause multipaction discharges in the devices (Addamo et al., 2010). This phenomenon is an exponential growth of electrons emitted by the metallic surfaces due to the synchronism between the applied electromagnetic field and the free electrons inside the components. The final effect consists in the damage and even in the destruction of the RF device.

2. Antenna-feed system architecture

The most general architecture of a dual-band dual-polarization antenna-feed system is shown in Fig. 2.1, where the paths covered by both the transmitter signals (in blue) and the receiver signals (in red) are reported. The same antenna is employed in both the transmitting (Tx) and receiving (Rx) mode, since the transmitters and the receivers works on different frequency bands (Harwanger et al., 2007), (Cecchini et al., 2009). By considering the system in the transmitting mode, the two separate microwave sources generate two independent signals. These are combined by the ortho-mode transducer (OMT) (Peverini et al., 2009) to obtain two orthogonal linear polarizations in a common waveguide connected to the antenna-feed. If circular polarizations are required, then a polarizer is introduced between the OMT and the antenna-feed in order to convert the incoming linearly-polarized signals into left- and right-hand polarizations (Virone et al., 2008). Finally, the antenna-feed radiates the Tx signals onto the reflector system.

In the receiving mode, the two circularly-polarized signals collected by the feed are converted by the polarizer into two linear orthogonal ones. The OMT separates these two orthogonal linear polarizations by routing them into two different single-polarization channels. Subsequently, the signals are amplified by low-noise amplifiers (LNAs), and elaborated by the back-end electronic of the receiver. Since the Rx and Tx signals are allocated in different frequency bands, they can coexist in the first stages of the chain without interfering each other. The diplexers allow the separation of these signals by routing them in different paths as a function of the frequency (Virone et al., 2009). In order to protect the receivers from spurious interfering signals generated by both internal and external transmitters operating in different frequency bands, various stop-band filters are inserted in the chain. Moreover, the correct behaviour of this system requires high performances also in terms of polarization purity. For this purpose, usually a corrugated horn is employed as the antenna-feed for its significant performances in terms of wide band and low cross-polarization levels (Addamo et al., 2009), (Beniguel et al., 2005).

An overview on the passive waveguide devices composing this chain (i.e. filters, diplexers, OMTs, and feed-horns) is reported in the next sessions, focusing on the main issues and characteristics (i.e. power-handling capability), and on the design techniques.

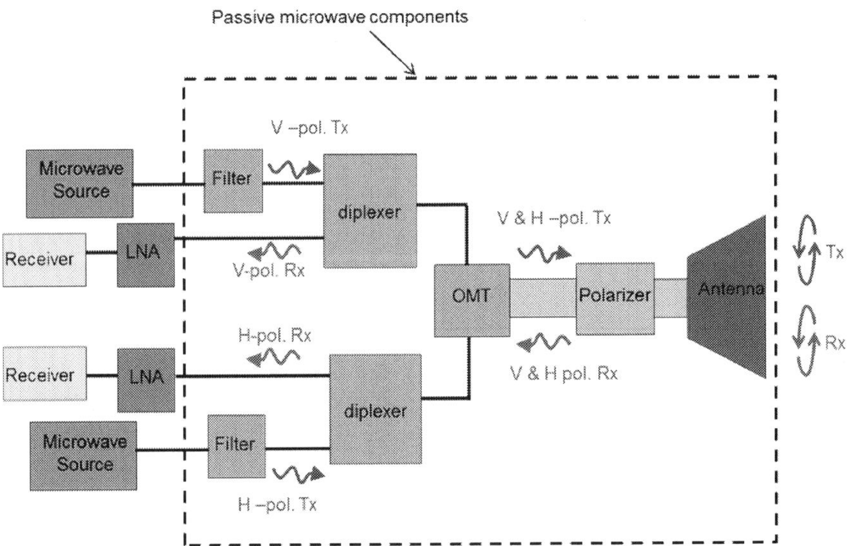

Fig. 2.1. General architecture of a dual-band dual-polarization antenna-feed system for satellite communication.

3. Multipactor discharge and passive intermodulation products

The trend in modern communication satellites is to increase the number of channels that can be handled by each RF payload in order to both minimize the mass, volume, and cost of satellites and to increase the reconfigurabilty flexibility of these communication systems. As an example, dual-band and tri-band RF chains operating in multi-carrier condition are currently adopted in several satellite programs for broadcast and fixed services in Ku, K, Ka-bands (Cecchini, et al., 2010). Consequently, the RF peak-power supplied to the antenna feed-systems can reach tens of kW. When such levels of power are employed, spurious unwanted phenomena can occur inside the components that can concretely damage or limit the operative function of the entire antenna-feed system. The most considerable ones are the multipactor discharge phenomenon and the generation of passive intermodulation products (PIMPs).

3.1 Multipactor discarge

Multipactor discharge is a breakdown mechanism that can occur under high-power and low-pressure conditions when a proper synchronism condition between the applied electromagnetic (EM) field and free electrons inside the components is met. Indeed, free electrons in microwave devices operating at low-pressure conditions can be accelerated by the applied EM field and impact onto the metallic internal surfaces. This impact can generate additional electrons by secondary emission that in turn can strike other surfaces. If appropriate dynamical conditions are satisfied, these repeated collisions and emissions can lead to an exponential growth of electrons and a subsequent discharge, thus increasing the noise level and modifying the electric parameters of the devices. Since the multipactor breakdown phenomenon sets severe constraints on the power level that can be handled by satellite payloads, specific device architectures are needed in order to overcome this problem. In particular, it is highly recommended that the design of each component satisfies suitable confidence margins with respect to multipaction discharge so that time- and cost-expensive experimental high-power testing can be avoided in the qualification process of the payloads. The multipactor breakdown phenomenon is based on a resonance condition that can arise when the electron mean free path (i.e. the average distance covered by a moving particle between successive collisions) is greater than the distance between two opposite metallic surfaces. Moreover, two additional conditions are needed. The first one is related to the impact time between two subsequent collisions onto the metallic surfaces. Under a single-carrier condition, this parameter has to be an odd number M of half cycles of the applied EM signal. The second condition implies that the arrival electron energy onto the metallic surface is sufficiently considerable so that the effective secondary emission ratio δ is greater than one. The latter coefficient depends also on the electrons incident angle, the surface material (typically aluminum), and the coating process applied to the metallic surfaces (e.g. silver-plating, alodine coating).

In order to gain a physical insight into the multipactor discharge phenomenon, it is useful to consider the simple model of a free electron in a plane parallel-plate waveguide where an electric field with TEM modal voltage $V(t) = V_0 \sin \omega t$ is present (see Fig. 3.1). The motion equation for an electron of mass m and charge e is

$$m\ddot{x}(t) = \frac{e}{d} V_0 \sin \omega t$$

(1)

together with the initial condition $\dot{x}(t_0) = v_0$ and $x(t_0) = 0$. If an electron is released from surface $x=0$ with velocity v_0, then integration of Eq. (1) yields the velocity from which the position can be derived as

$$\dot{x}(t) = \frac{e}{md}\frac{V_0}{\omega}\left(\cos\alpha - \cos\omega t\right) + v_0 \tag{2}$$

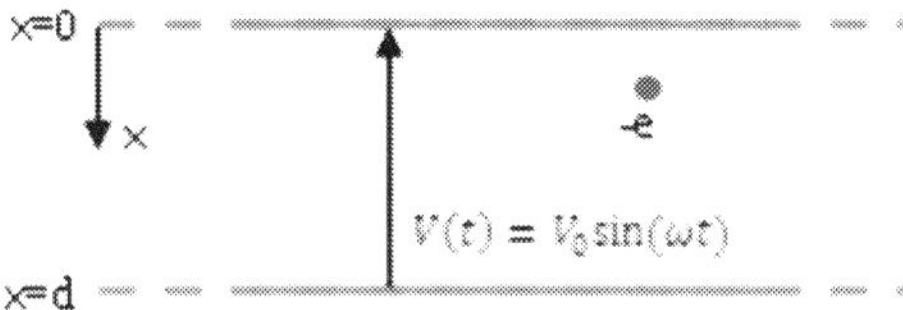

Fig. 3.1. Plane parallel-plate waveguide under consideration for the analysis of the multipaction breakdown phenomenon.

$$x(t) = \frac{e}{md}\frac{V_0}{\omega^2}\left((\omega t - \alpha)\cos\alpha - \sin\omega t + \sin\alpha\right) + \frac{v_0}{\omega}(\omega t - \alpha) \tag{3}$$

where $\alpha = \omega t_0$. In order the transit time Δt to the position $x=d$ to be an odd number M of half cycles (i.e. $\Delta t = M\pi / \omega$), the peak voltage V_0 of the EM field and the electron impact velocity v_i must be equal to

$$V_0 = \frac{m}{e}\frac{\omega d(\omega d - M\pi v_0)}{M\pi\cos\alpha + 2\sin\alpha} \tag{4}$$

$$v_i = \frac{2e}{m}\frac{V_0}{\omega d}\cos\alpha + v_0 \tag{5}$$

respectively. According to Eq. (4) the value of the peak voltage V_0 fulfilling the synchronism condition (named also the multipactor threshold voltage) depends on α (i.e. the time t_0 when the electrons are released from the surface $x=0$) and on the gap-frequency product (i.e. fd). With reference to Eq. (5), it is worth noting that the kinetic energy $\frac{mv_i^2}{2}$ of the primary electron striking the surface $x=d$ depends on the peak voltage V_0. If this energy is sufficiently high so that the secondary electron emission coefficient δ of the surfaces is greater than one, than an electron avalanche phenomenon occurs between the two surfaces. On the basis of the previous theory, it is possible to derive the relationship between the gap-frequency product fd and the threshold voltage V_0 for each odd-order resonance, thus obtaining useful design tools such as the Hatch-Williams susceptibility diagrams. The latter are the basis of the free multipactor calculator program developed by the European Space Agency and available online (ESA, 2007), which provides the region in the $fd - V_0$ plane where the multipaction breakdown can occur for a given resonance order. Fig. 3.2 shows the envelope over all the resonance orders of the minimum threshold voltage for a silver-plated parallel-plate waveguide. It is worth mentioning that since the EM field inside a generic component can deviate significantly from the TEM field in a plane parallel-plate waveguide, the rigorous analysis of the multipaction breakdown would require extensive numerical computations of the electronic trajectories inside the devices in order to establish if an avalanche of secondary-emission electrons can occur (Anza, et al., 2008)-(Tienda, et al, 2006).

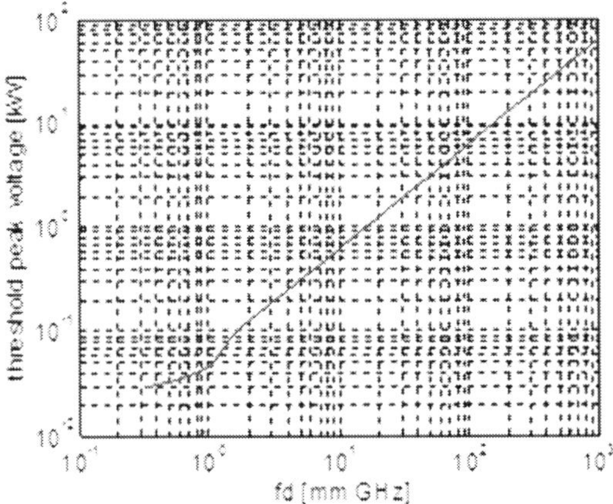

Fig. 3.2. Envelope over all the resonance order of the minimum multipactor threshold voltage V_0 as a function of the gap-frequency product fd for a silver-plated parallel-plate waveguide.

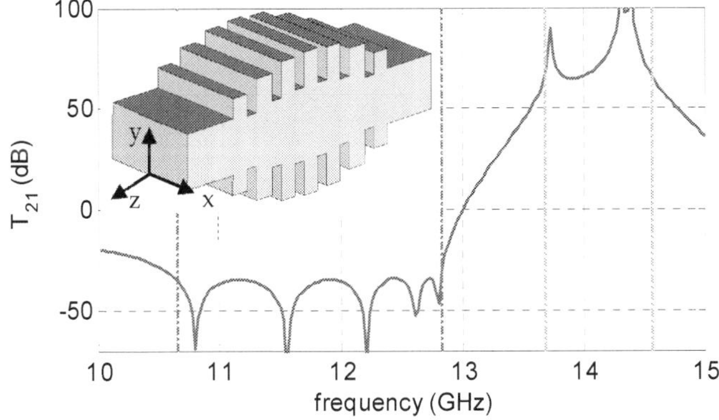

Fig. 3.3. Transmission coefficient $T_{21}(f)$ of the E-plane WR75-waveguide symmetric stub-filter shown in the insert (the inside waveguide structure of the filter is reported).

However, according to the ESA recommendations, the qualification process of a generic RF component in terms of the power-handling under both single- and multi-carrier operating conditions can be carried out by evaluating an upper-bound on the multipaction risk and setting appropriate confidence margins. In particular, the actual upper-bound is computed by using the plane parallel-plate model along several directions inside the component. For sake of clarity and without loss of generality, this procedure is described next by referring to the E-plane WR75-waveguide symmetric stub-filter depicted in the insert of Fig. 3.3, where the transmission coefficient $T_{21}(f)$ of this filter is reported. The transmission coefficient is the relevant characteristic function of the filter, since it is equal to the ratio $S_{11}(f) / S_{21}(f)$ where $S_{11}(f)$ is the scattering reflection coefficient at the input port, and $S_{21}(f)$ is the scattering transmission coefficient from one port to the other. Hence, the transmission coefficient $T_{21}(f)$ is proportional to the reflection coefficient in the pass-band and to the inverse of the transmission between the two ports in the stop-band (isolation). The E-plane stub architecture is commonly adopted in the Tx-channels of multiplexers (Tx-band = [10.7, 12.75] GHz) to block the Rx signals (Rx-band = [13.5, 14.5] GHz), since each stub exhibits a transmission zero that can be adjusted in the stop-band by varying its length. In this way, high levels of isolation can be achieved in the Rx-band along with very low standing-wave ratio inside the component in the Tx-band. The latter condition can be exploited in order to maximize the power-handling capability of these components.

Since this filter is an E-plane structure, the maximum electric field arises in the central plane x=0, for which the in-phase field lines at 13 GHz are depicted in Fig. 3.4. Although the field in the device is not everywhere oriented along straight lines connecting two parallel surfaces (as in the parallel-plate model), it is possible to define a parallel-plate model for each of the lines highlighted in cyan in Fig. 3.4. For this propose, the equivalent voltage

$$V_i(f) = \int_0^{d_1} \underline{E}(s;f) \cdot \hat{s}\, ds \tag{6}$$

is evaluated on the i-th integration line (oriented along $\hat{s}$). Moreover, the corresponding multipaction threshold voltage $V_i^{(thres)}$ for this section of the device, can be evaluated in terms of the frequency-gap product fd_i by means of the susceptibility diagrams. For design purposes, it is useful to introduce the voltage magnification factor VMF_i (Parikh, et al., 2003) that provides a measure of the magnification of the electrical field occurring in the i-th position referred to the incident voltage $V^{(inc)}$

$$VMF_i(f) = \frac{V_i(f)}{V^{(inc)}} \tag{7}$$

Accordingly, a breakdown-free condition is guaranteed at the i-th section of the device if the input power is smaller than the threshold level

$$P_i(f) = \frac{|V_i^{(thres)}|^2}{2Z^{(inc)}|VMF_i|^2} \tag{8}$$

where $Z^{(inc)}$ is the power-voltage impedance at the input waveguide port. Finally, the overall breakdown threshold power of the device at frequency f is

$$P^{(SC)}(f) = \min_i\{P_i(f)\} \tag{9}$$

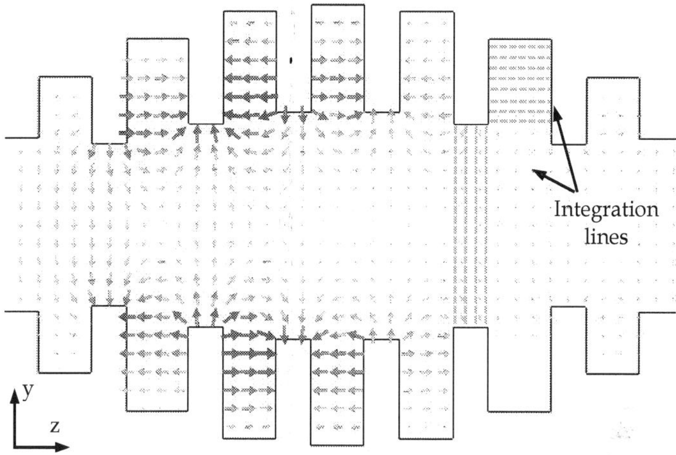

Fig. 3.4. In-phase field lines in the plane x=0 of the E-plane WR75-waveguide symmetric stub-filter shown in the insert of Fig. 3.3. The frequency is equal to 13 GHz. The lines highlighted in cyan correspond to the integration lines used to define the equivalent parallel-plate models.

that clearly defines the power-handling capability of the component in a single-carrier regime. Fig. 3.5 reports the frequency behavior of $P^{(SC)}(f)$ for the E-plane stub-filter in the Tx-band. The minimum value of 2.6 kW at 12.75 GHz is due to the high levels of standing-waves that are established inside the stubs at frequencies close to the -3 dB cut-off frequency, in order to achieve high levels of isolation in the stop-band [13.5, 14.5] GHz. In this regard, the power-handling capability of any device can be increased by adopting the following strategies:

- Enlargement of the design bandwidth with respect to the actual operating bandwidth of the device. In this way, as stated previously, the power-handling capability is not adversely affected by very high standing-waves inside the component towards the band limits.

- Application of surface-coating processes (i.e. silver-plating), since they guarantee higher breakdown threshold voltages with respect to bare aluminum. It is worth mentioning that the choice of the specific surface treatment has to be made by considering both the insertion loss and the power-handling requirements.

- Setting proper constraints on the geometric parameters during the design of the architecture. Indeed, a significant improvement in the power-handling capability can be achieved by varying the height of the most critical sections of the component under analysis. This leads to a larger frequency-gap product and, consequently, to higher value of breakdown voltages. Hence, the geometrical parameters of the architecture are determined through a trade-off process between the electrical requirements (e.g. return-loss at the input ports or channel isolation) and the power-handling capability of the device. In this view, the design of novel instrumentation architectures exhibiting very good electrical figure-of-merits along with very high power-handling capabilities is a cutting-edge research topic for satellite communication systems.

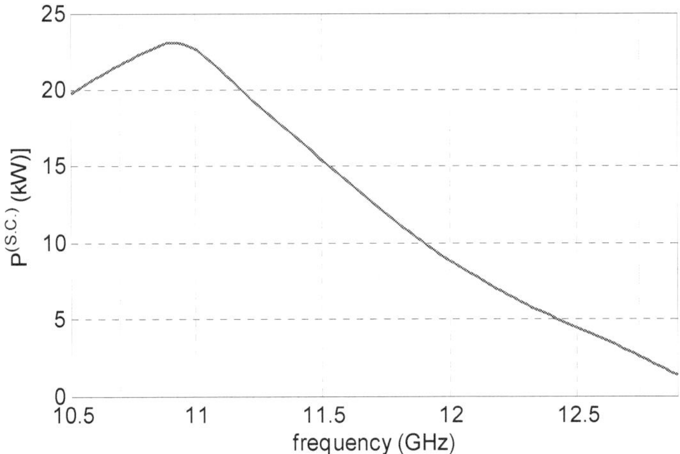

Fig. 3.5. Single-carrier breakdown threshold power $P^{(SC)}(f)$ for the E-plane WR75-waveguide symmetric stub-filter shown in the insert of Fig. 3.3.

On the basis of the single-carrier analysis previously described, it is possible to derive the relevant design upper-bounds on the maximum power deliverable to the device operating in a multi-carrier condition. Under the assumption of N carriers with equal power P, the worst case corresponds to the in-phase sum of the carrier fields, thus leading to a total peak-power equal to $N^2 P$. As a consequence, the breakdown-free condition in the device is certainly guaranteed if the input power per carrier P is smaller than the threshold level

$$P^{(MC)} = \frac{1}{N^2} \min_f \{ P^{(MC)}(f) \} \tag{10}$$

By considering a further margin of 3 or 6 dB, the standardized $"N^2 \cdot P + 6\,dB"$ or $"N^2 \cdot P + 3\,dB"$ rules are derived. Actually, these upper-bounds provide to be too strict when a high number of carriers are considered. Indeed, the in-phase condition of the N carriers can be satisfied only for a short span of time. Moreover, the multipaction breakdown is an electron secondary-emission resonance that has to be sustained by the applied EM field. For these reasons, the in-phase matching condition becomes critical for the multipactor breakdown only if it is satisfied for long time scales. In this respect, the high-power qualification process of the devices operating in a multi-carrier regime is usually carried out by adopting the more realistic "20-gap crossing" rule. The latter states that "as long as the duration of the multi-carrier peak and the mode order gap are such that no more than 20 gap-crossings can occur during the multi-carrier peak, the design may be considered safe with regards to the multipaction breakdown even though the multipaction threshold may be exceed from time to time". Implementation of this rule in the case of N linearly-spaced carriers (frequency spacing Δf) yields the definition of the boundary function (Parikh, et al., 2003).

$$F_V = -\left(\sqrt{N}-1\right)\ln\left(\frac{T_{20}}{T_H}\right)+\sqrt{N} \qquad for \qquad \sqrt{N} < F_V < N \tag{11}$$

where T_H is the period of multi-carrier envelope ($T_H = 1 / \Delta f$) and T_{20} is the time taken by the electrons to cross the most critical gap 20 times. The latter parameter is equal to $T_{20} = (20 \times n) / (2 \times f_0)$, where n is the resonance order fulfilling the synchronism condition and f_0 is the lower frequency in the band of interest. On the basis of this boundary function, the maximum power per carrier satisfying to the "20-gap crossing " is

$$P^{(MC)} = \frac{1}{F_V^2} \, mi \, \underset{f}{n} \{P^{(MC)}(f)\} \tag{12}$$

to which a further 6 dB confidence margin is commonly added, thus defining the "20-gap crossing + 6 dB" rule. As an example, when considering 10 carriers linearly spaced in the Tx-band of the E-plane stub filter, the maximum power per carrier according to this rule is approximately 67 W.

3.2 Passive intermodulation products

Nonlinear characteristics in microwave components can lead to the generation of spurious passive intermodulation products (PIMPs). When the intermodulation products of two or more signals mixed in the device fall into the operative bandwidth of the receiver, this intermodulation signal becomes an interference problem (Lui, 1990). As an example, if two carriers with frequencies f_1 and f_2 propagate through a nonlinear passive component, the spurious intermodulation products are harmonics with frequencies $f_{m,n} = mf_1 + nf_2$ with m, n integers. The sum $|m|+|n|$ defines the order of the intermodulation product and the amplitude of the PIMPs rapidly decay as a function of the order $|m|+|n|$. However, for the case of considerable input power, some of the higher-order products can be great enough to cause serious interference problems. This usually happens in satellite communication systems where high-power transmitters and low-noise receivers are employed in the same antenna-feed system. As a consequence, appropriate counter-measures have to be taken in order to avoid the decrease of the signal-to-noise ratio in the Rx channels, which in turn reduces the receivers sensibility. As an example, PIMPs level as low as -140 dB are commonly required in Ku, K, Ka-band payloads operating broadcast and fixed satellite services.

Generation of PIMPs take place mainly in the Tx power-amplifier circuits, in the receiver mixers, and in the nonlinear metallic contacts inside the antenna-feed systems. The effects of PIMPs generated in the back-end circuits (amplifiers and mixers) can be minimized by inserting ad-hoc filters. On the contrary, PIMPs generated by possible metallic-oxide-metallic contacts arising in the metallic mating surfaces of the front-end system components are more troublesome. Indeed, depending on the specific position of the intermodulation surface inside the antenna-feed chain, PIMPs can even not be filtered out. In this regard, the level of PIMPs generated in an oxidized surface that mates two metallic blocks depends significantly on the current through the junction. For this reason, the electrical and mechanical designs of all the front-end components are strictly connected. Indeed, special attention has to be paid when splitting a component in several blocks and in the connection of the components.

With regards to the E- plane stub-filter described in Sec. 3.1, the clam-shell assembly shown in Fig. 3.6 is a mechanical implementation of this device that is optimized In terms of PIMPs generation. The device is halved in two blocks along the central plane $x=0$, thus allowing a milling manufacturing of the inside waveguide structure. Since the currents in the central plane x=0 are oriented along the longitudinal z-direction, no currents cross the two mating surfaces, thus avoiding the generation of PIMPs. Finally, the PIMPs generated at the input port sections, where the filter is connected to the other components, are minimized by adopting a choke/plain joint consisting of a choke flange (applied to the filter) and a plain flange (applied to the connecting device).

Fig. 3.7 shows the contour plot of the magnetic field amplitude inside the choke/plain joint at 12.75 GHz. It is worth noting, that the magnetic field, hence the electric current, in the contact point between the two flanges (named also cold point) is minimized with an appropriate design of the resulting L-shaped radial stub. Moreover, the joint is designed to exhibit a return-loss as high as possible in the operating bands (as high as 40 dB).

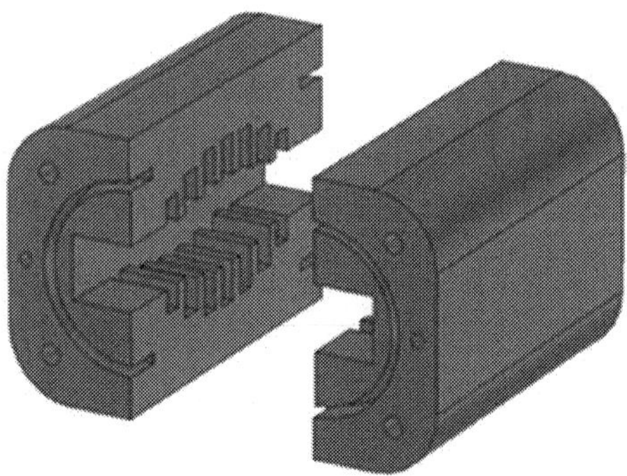

Fig. 3.6. Clam-shell mechanical assembly of the E-plane stub-filter shown in the insert of Fig. 3.3.

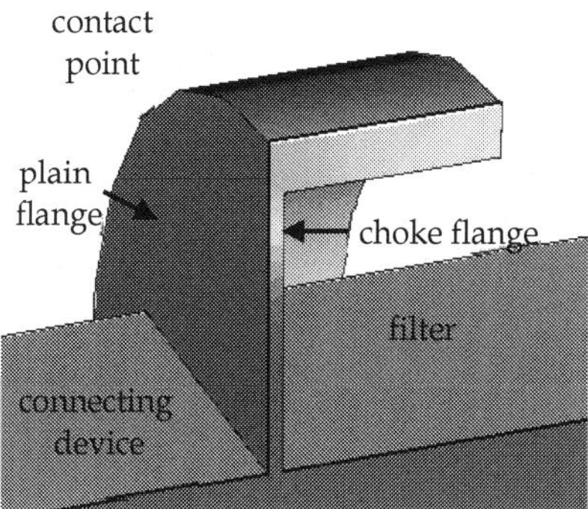

Fig. 3.7. Contour plot of the magnetic field inside the L-shaped radial stub resulting from the connection of the choke and plain flanges used to mount the filter of Fig. 3.6 with a connecting device (a rainbow contour scale is used).

4. Broadband waveguide filters and diplexers

Metal waveguide filters are typically employed in satellite antenna-feed systems for their low losses and high power-handling at the microwave frequencies. As discussed in the introduction, these structures are mainly used to separate different sub-bands e.g. receive and transmit bands as well as to protect the source from spurious signals. The latter operation is usually performed using a single pass-band filter. The sub-band separation is instead performed using two (or more) filters in the diplexer (multiplexer) configuration. The same operation could be performed using a circulator, however, the diplexer solution exhibits high-performance and a low-cost.

A general diplexer configuration is sketched in Fig. 4.1, where two different filters (TX and RX) are connected to a three-port (T or Y) junction in order to obtain a common port (Port 1). The other filter ports are instead connected to proper waveguide transitions to provide the required orientation and size of Ports 2 and 3. More complex junctions could be adopted at port 1 in order to increase the number of sub-bands.

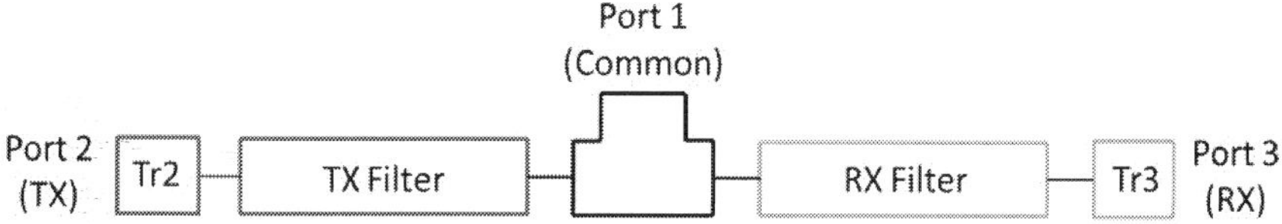

Fig. 4.1. Scheme of a waveguide diplexer.

With reference to the diplexer architecture in Fig. 4.1, the basic electrical requirements are a high transmission coefficient from Port 2 to Port 1, high attenuation from Port 2 to Port 3 and a low reflection coefficient at Port 2 in the TX frequency band. A high-transmission coefficient from Port 1 to Port 3, a high attenuation from Port 1 to Port 2, and a low-reflection coefficient at port 3 have to be instead provided in the RX band. A low reflection coefficient at Port 1 for both frequency bands is also required.

It should be pointed out that filtering structures with relatively broad pass-bands (more than 5-10 %) are required owing to the present specifications of the satellite antenna feed systems. For this reason, specific synthesis techniques based on distributed parameter models and full-wave analysis tools should be adopted to design these kind of filters. These filters and their corresponding design procedures are hence very different with respect to narrow band (0.2-0.3 %) channel filters (not treated in this section) where the frequency dispersion of the discontinuities around the pass-band is practically negligible.

The filters for the antenna feed system diplexers can be designed according to either the pass-band or the stop-band architecture. Both of them can in principle be represented with the fundamental-mode equivalent circuit of Fig. 4.2.

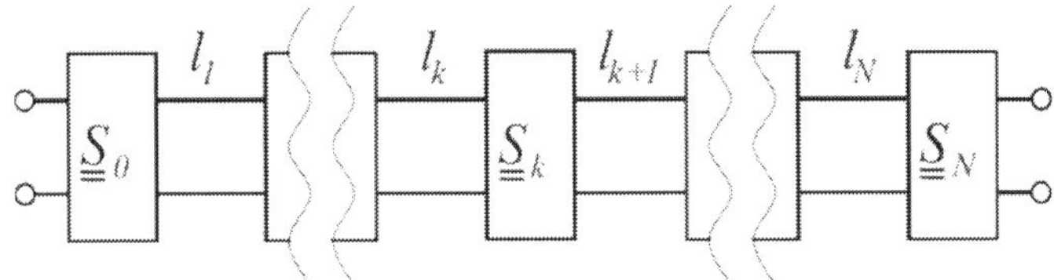

Fig. 4.2. Fundamental-mode transmission line equivalent circuit of a waveguide filter.

Such a circuit consists of N+1 scattering matrices $\underline{S}_k$, with $k = 0, \cdots, N$, connected by N transmission lines representing the same number of generic waveguide discontinuities and waveguide sections, respectively. The parameter l_k defines the length of these sections.

In pass-band architectures, the filtering behavior is mainly related to the phase rotation versus frequency in the N waveguide sections. In this framework, the latter are in fact usually referred as cavities or resonators. The main role of the discontinuities is instead to provide the required coupling between the adjacent resonators. However, as it will be discussed in the following, the spurious dispersive effect of the various discontinuities significantly affects the overall frequency behavior of the filter. Therefore, it should be kept into account in the design stage.

As far the stop-band architecture is concerned, the required transmission zeros are introduced by the discontinuities themselves, which exhibit a strong resonant behavior in this case. The spacing between the various discontinuities is instead adjusted to obtain a good matching in the pass-band.

The correct choice between the two architectures mainly depends on the overall required frequency behavior i.e. the width of the pass-, stop- and transition bands, the power handling capability, losses and the manufacturing complexity. Both the architectures will be discussed in the remainder of this section.

4.1 Pass-band structures

Generally speaking, two class of discontinuities can be adopted in the design of pass-band filters. The first one is represented by the transverse discontinuities i.e. inductive (Rozzi, 1972) or capacitive irises (Virone, et al. 2007). A band-pass configuration with inductive (or

H-plane) irises is shown in Fig. 4.3. As it can be seen, five resonators in rectangular waveguide are obtained owing to the presence of the six inductive irises. The peculiarity of this structure is the increased reflection coefficient of the irises at lower frequencies which lead to very high attenuation levels in the frequency region below the pass-band. The opposite phenomenon occurs with the capacitive configuration shown in Fig. 4.4. Indeed, the reflection coefficient of capacitive irises increases at higher frequency providing a very high attenuation above the pass-band of the whole filter. It has to be pointed out that waveguide resonators with an increased height (see Fig. 4.4) are used to reduce the overall losses.

For both the capacitive and inductive configurations, iris apertures and resonator lengths are the main design parameters. The iris thickness is generally selected according to the manufacturing materials and techniques. In particular, proper rounding of some of the filter corners is also required when milling machines are adopted. Nevertheless, this feature can be kept into account in modern design tools (Arndt, et al. 1997) in order to avoid the insertion of tuning screws.

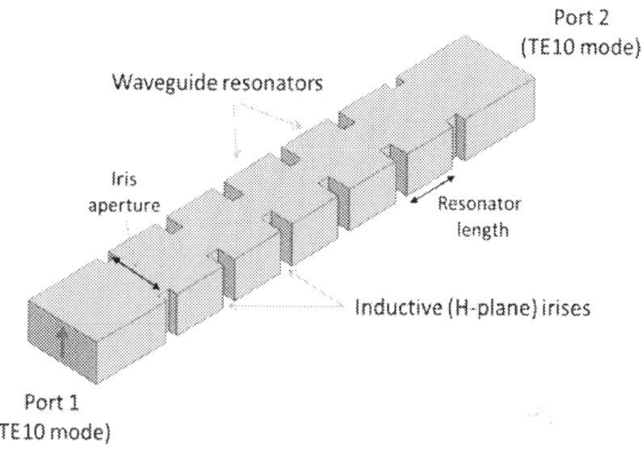

Fig. 4.3. Pass-band filter configuration with inductive (*H*-plane) irises.

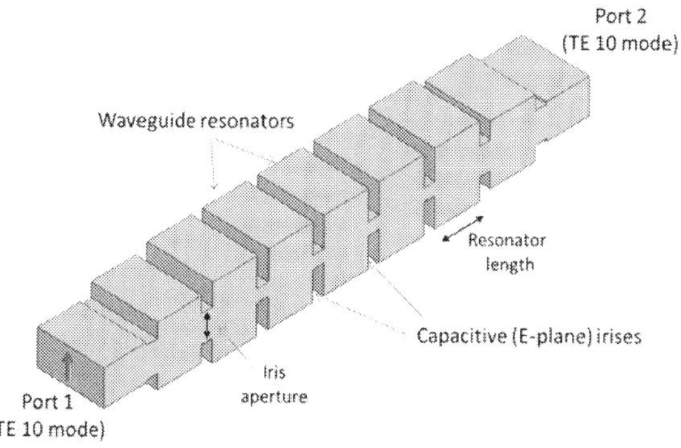

Fig. 4.4. Pass-band filter configuration with capacitive (*E*-plane) irises.

The second class of discontinuities for pass-band filters is represented by the longitudinal ones. Among these, the *E*-plane septum configuration shown in Fig. 4.5 is very popular (Vahldieck, et al, 1983). Such a discontinuity provides a very high reflection coefficient because the septum is placed in the middle of the waveguide where the electric field is maximum. Moreover, the electromagnetic field is evanescent in the septum region owing to the splitting of the main rectangular waveguide in two halves for which the TE10 is below-cut off. The design parameters of the septum filter are both the resonator lengths and septum lengths. The septum width is usually selected according to manufacturing considerations. It should be pointed out that the septum reflection coefficient can even be too high for certain broadband applications. Therefore, open septa can be adopted as first and last discontinuities (Peverini, 2004). More advanced configurations feature ridge waveguide resonators, instead of the common rectangular ones, in order to decrease the overall length of the filter (Goussetis and Budimir, 2001).

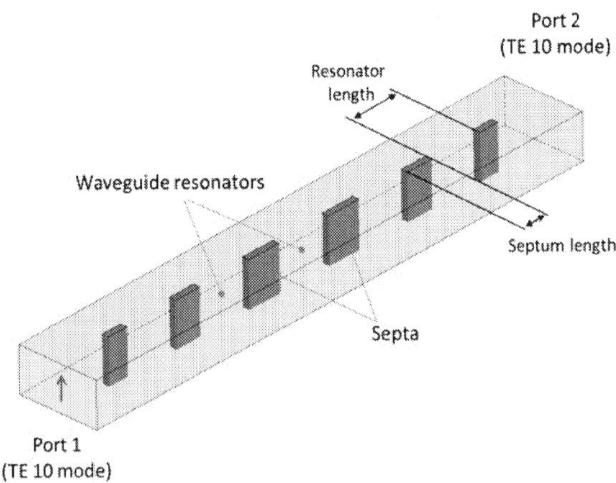

Fig. 4.5. Pass-band filter configuration with *E*-plane septum discontinuities.

The evanescent mode filter is another common structure featuring longitudinal discontinuities (Bornemann and Arndt, et al. 1990). As shown in Fig. 4.6, this configuration is based on a dual ridge waveguide (single ridge versions are also used). Therefore, it leads to more compact implementations in terms of both length and transverse section with respect to the rectangular counterparts. The smaller transverse section also produce a wider attenuation bandwidth. The small gap between the two ridges however generally reduce the power handling of the structure owing to the multipactor phenomenon. The longitudinal discontinuity is represented by the interruption of the ridge. In particular, the envelope of the adopted ridge waveguide is selected so that the TE10 mode in the discontinuity region is far below cut-off in the operative frequency band. In this way, a strong evanescent-mode discontinuity is created. Besides the dimensions of the ridge waveguide, the relevant parameters for the filter design are the lengths of both the resonators and the evanescent mode sections.

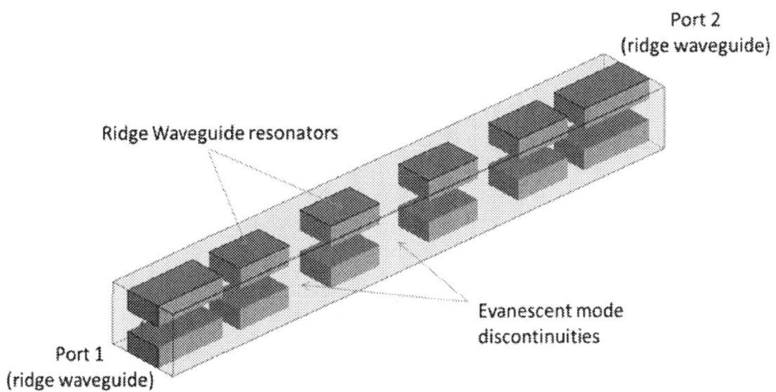

Fig. 4.6. Evanescent mode filter.

As discussed above, all the filter geometries exhibit a large number of design parameters that should be defined according to the required specifications on the frequency response. In the past, this task has been generally accomplished on the basis of synthesis techniques for lumped-parameters circuits and proper transformations (Levy, 1984). However, these techniques do not take into account the frequency dispersion of both the discontinuities and the cavities, the multimodal interactions and the losses which can significantly affect the frequency response of a broadband filter. As a consequence, a considerable optimization work is required to refine the initial solution. In particular, such an approach could lead to sub-optimum designs owing to the presence of local minima in the pertaining cost function.

In order to overcome the above-mentioned problems, a design method based on both a distributed-parameter model of the structure and a proper identification technique for the various higher-order spurious effects has been developed (Tascone, et al. 2000). The method exploits the single-mode circuit shown in Fig. 4.2, where each discontinuity is described using its scattering parameters $S_{11}^{(k)}, S_{21}^{(k)}$ and $S_{22}^{(k)}$. In the lossless case, these complex quantities can be conveniently represented using the real parameter γ_k ($\cos\gamma_k = |S_{11}^{(k)}| = |S_{22}^{(k)}|$ and $\sin\gamma_k = |S_{21}^{(k)}|$) and the phases $\varphi_{11}^{(k)}$ and $\varphi_{22}^{(k)}$. The resonators are instead described using their lengths l_k and the corresponding waveguide propagation constants β_k. It has been demonstrated that the transmission matrix of the complete circuit can be written as

$$T = \left\{ \prod_{k=1,\cdots,n} \begin{bmatrix} \csc\gamma_k & \cot\gamma_k z^{-1}e^{-j\psi_k} \\ \cot\gamma_k & \csc\gamma_k z^{-1}e^{-j\psi_k} \end{bmatrix} \right\} \begin{bmatrix} \csc\gamma_0 & \cot\gamma_0 \\ \cot\gamma_0 & \csc\gamma_k \end{bmatrix}$$

where $z^{-1}e^{-j\psi_k} = \exp\{-j[2\beta_k l_k - (\varphi_{11}^{(k-1)} + \varphi_{22}^{(k)})]\}$ and ψ_k are unknown phase terms which have been introduced in order to allow the synthesis of arbitrary frequency responses and to keep the different phase behavior of the resonators into account. It should be noted that the quantity $2\beta_k l_k - (\varphi_{11}^{(k-1)} + \varphi_{22}^{(k)})$ is the Round Trip Phase Shift of the resonators. It can be also recognized that the elements of the transmission matrix T are N-degree polynomials in the complex variable z^{-1}

$$T_{11}(z) = \sum_{k=1}^{N} a_k z^{-k} \text{ and } T_{21}(z) = \sum_{k=1}^{N} b_k z^{-k}$$

As well-known, these parameters are related to the scattering matrix of the overall filter

$$T_{11} = 1 / S_{11} \text{ and } T_{21} = S_{11} / S_{21}$$

In particular, the T_{21} parameter provides a very convenient description of the frequency response in both the passband (where $T_{21} \approx S_{11}$) and the attenuated bands (where $T_{21} \approx 1 / S_{21}$). Thanks to the polynomial representation, this single parameter can be used to analytically define all the specifications of the filter in the same fashion as antenna arrays and digital filter design techniques (FIR).

Once the polynomial T_{21} has been defined according to the required specifications, an extraction procedure is applied to determine the scattering matrix of the various discontinuities (Tascone, et al. 2000) and the phase terms ψ_k. Starting from the electrical parameters, the geometry of the discontinuities e.g. iris apertures, septum lengths are obtained exploiting a pre-computed curve. The length of the various resonators l_k is also computed exploiting both the ψ_k and the $\varphi_{11}^{(k-1)} + \varphi_{22}^{(k)}$.

The synthesized geometry is then simulated using a full-wave method (e.g. Peverini, 2004). As expected, the computed full-wave response is different from the synthesized polynomial T_{21} owing to the neglected phenomena in the synthesis model. Nevertheless, these discrepancies can be properly compensated using the developed identification procedure (Tascone, et al. 2000). Such a technique starts with the interpolation of the full-wave response with a N-degree polynomial

$$A_{21}(z) = \sum_{k=1}^{N} c_k z^{-k}$$

In this way, the overall procedure from the synthesis scheme to the analysis tool can be represented as a single abstract system having the $T_{21}(z)$ and $A_{21}(z)$ polynomials as input and output, respectively. These quantities can be interpreted as periodic signals with N harmonics. Hence, their coefficients b_k and c_k can be considered as the discrete spectra of these input and output signals, respectively. This approach leads to identification of the above mentioned abstract system with a linear system, for which the transfer function $H(z)$ can be defined as the ratio between the output and input spectra

$$h_k = c_k / b_k$$

The introduced transfer function $H(z)$ properly models the frequency dispersion of the discontinuities, the multimodal interaction and the losses. Therefore, it can be used to compute the coefficients a new "pre-distorted" frequency response $T_{21}'(z)$ as

$$b_k' = b_k / h_k$$

Finally, the synthesis scheme discussed above is applied to this "pre-distorted" polynomial $h_k = c_k / b_k$ to obtain the final filter geometry. Indeed, a single iteration is usually enough for several considered filter designs. Anyway, some more iterations can be performed in presence of highly-dispersive discontinuities or strong coupling to the evanescent modes.

The presented synthesis technique has been generalized to the design of waveguide polarizers (Virone, et al. 2005) and waveguide diplexers (Virone, et al. 2009). For the latter, the frequency response of a relevant design example for Digital Video Broadcasting applications is reported in Fig. 4.7. It has been conceived with the capacitive iris configuration (see Fig. 4.4). With reference to the port numbering reported in Fig. 4.1, the reflection coefficient at common port (S_{11}) is below -30 dB for both the TX (10.95-12.75 GHz) and RX (13.75-14.5 GHz) frequency bands. Similar reflection levels have been obtained at both port 2 and 3 for the TX and RX bands, respectively. Due to the metallic losses, the transmission coefficient is about -0.4 dB in the pass-band of both channels (an aluminum prototype has been simulated). The isolation levels $1/S_{21}$ (from TX to port 1) and $1/S_{31}$ (from 1 to RX) are higher than 55 dB in the RX and TX band, respectively. A level of 55 dB has been obtained in both bands from TX to RX ($1/S_{32}$) as well.

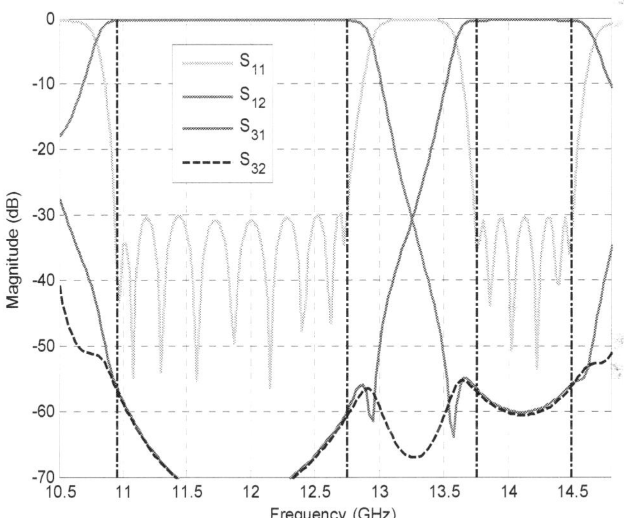

Fig. 4.7. Waveguide diplexer with capacitive irises. The port numbering is reported in Fig. 4.1. The vertical dash-dotted lines represent the TX (10.95-12.75 GHz) and RX (13.75-14.5 GHz) frequency bands.

4.2 Stop-band structures

Strongly resonant waveguide discontinuities are required to design a stop-band filter. In particular, they have to exhibit both a relatively wide transmission zero around the specified stop-band and a moderate reflection level in the pass-band. A very common configuration is the E-plane stub in rectangular waveguide shown in Figs. 3.3-3.4. Several discontinuities have to be cascaded to meet the specifications. The number of stubs is mainly related to the desired isolation level in the given bandwidth. Proper arranging of the various discontinuities is mandatory to avoid degradation of the stop-band performance owing to tunnel phenomena and strong multimode interactions. A significant design example of stub-filter for satellite applications has already been described in section 3. Other designs can be found in (Kirilenko, et al. 1994) and (Levy, 2009).

5. Ortho-Mode Transducers

A canonical Ortho-Mode Transducer (OMT) configuration has three physical waveguide ports (see Fig. 5.1). The common port is a metallic waveguide having a two-fold symmetric cross section (e.g. square or circular) which supports two degenerate modes i.e. two orthogonal polarizations. Hence, two electrical ports should in fact be adopted to describe the electromagnetic field at the common port. With reference to Fig. 5.1, the symbols V (H) is hereinafter used to indicate the vertical (horizontal) polarization and Port 1 (2) is referred as the electrical port of the component V (H). The two coupled ports are usually rectangular waveguides operating in the single-mode regime. In other words, the OMT should in fact be described using the four-port scattering matrix shown in Fig. 5.1 (Peverini, et al. 2006).

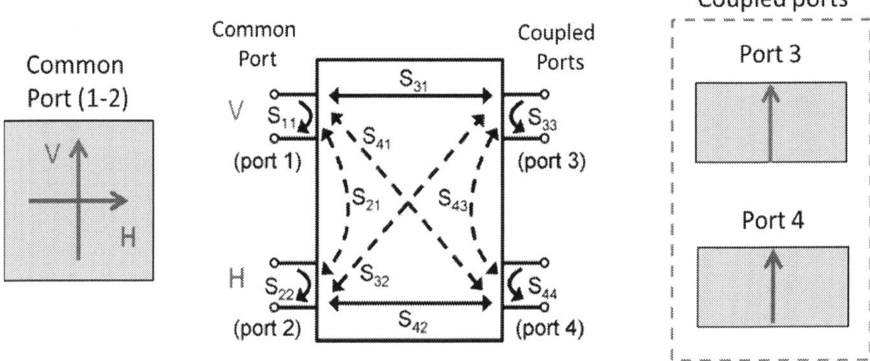

Fig. 5.8. The three waveguide ports of the OMT and the four-port scattering matrix equivalent circuit.

An OMT waveguide architecture is conceived so that each of the two degenerate modes at the common port is nominally coupled to one of the two single-mode ports only. According to Fig. 5.1, Port 3 (4) is the coupled port of the component V (H). It means that the magnitude of the direct scattering parameter S_{31} (S_{42}) should be ideally equal to one. The magnitude of the cross-coupling terms S_{41} (S_{32}) and the isolation terms S_{21} and S_{43} should instead be zero.

According to a receive-mode operative condition, an OMT is therefore used to route the two orthogonally polarized field components at the common port to two different waveguide ports. By reciprocity, in transmit mode, an electrical signal excited at one of the two single-mode ports is only routed to one of the two available polarizations at the common port.

The routing of the various signals is obtained introducing suitable apertures and discontinuities between the common waveguide and the coupled waveguides. The isolation between the two signals pertaining to the orthogonal polarizations is generally obtained exploiting both the symmetry of the structure and particular cut-off phenomena.

Besides the isolation, other figures of merit of the OMT consist in power handling, insertion loss, matching level, and low spurious higher-order-mode excitation at the four electrical ports. Group delay equalization between the V and H channels is also required in some specific applications. All the relevant electrical parameters of the OMT should be robust with respect to the manufacturing tolerances. Manufacturing complexity, compactness and proper orientation of the physical ports are also very important design parameters.

Several OMT configurations can be found in the literature. Their performances can change significantly from both the electromagnetic and mechanical perspectives . The correct choice highly depends on the application, however, it is worth mentioning that this choice strongly affects the overall feed system topology. Therefore, it is important to correctly choose the OMT since it drives the complexity as well as the cost and the dimension of the overall antenna feed system.

The most interesting waveguide configurations for satellite communication systems will be discussed in the following, starting from the basic ones up to the more recent and advanced architectures. For the sake of simplicity, the square-waveguide common-port implementations are only shown i.e. the V (H) polarization shown in Fig. 5.1 corresponds to the TE10 (TE01) mode. Nevertheless, all the presented architectures can in principle be extended to the circular cases.

5.1 Standard T-junction or side-coupling OMT

One of the simplest OMT architectures is represented by the T-junction shown in Fig. 5.2 (Schlegel and Fowler, 1984). The TE10 mode incident at the common port (carrying the vertical polarization) propagates along the main arm (also called common waveguide) and it is coupled to the fundamental mode of the rectangular port 3. The TE10 in the common waveguide is not coupled to the fundamental TE10 at port 4 owing to both the symmetry of the structure and the orthogonality of their field distributions.

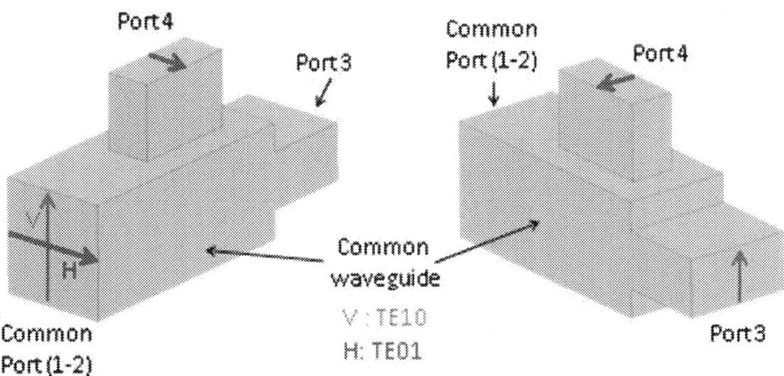

Fig. 5.9. Standard T-junction OMT.

The horizontal polarization, i.e. the TE01 mode at the common port is instead coupled to the side arm (fundamental mode of port 4) only. Indeed, the same polarization is under cut-off at port 3. As a consequence, ports 3 and 4 are also isolated as far as their fundamental mode is concerned.

A careful design of the various geometrical parameters is required in order to obtain an OMT with a suitable matching level. The side-arm coupling can be also performed on the other orthogonal side of the common waveguide with a different orientation of the coupled waveguide i.e. *E*-plane coupling instead of *H*-plane coupling. Anyway, this simple compact configuration only works in quite narrow frequency bands. Proper matching structures such as septa, irises and steps can be added to enlarge the operative frequency band up to 20% (Dunning, et al. 2009) or to obtain a dual-band component (Rebollar, 1998). However, proper care should be taken in order not to impair the power handling of the structure. Moreover, the bandwidth limit of this configuration is related to the excitation of the higher order modes TE11 and TM11 owing to the one-fold symmetry of the structure.

5.2 Boifot OMT

The Boifot junction has been introduced in order to obtain an OMT with a large operative bandwidth (Boifot, 1990). As can be seen in Fig. 5.3, a symmetric E-plane coupling is exploited for the horizontal polarization in order to obtain a two-fold symmetry of the whole structure. This feature avoids the excitation of the TE11 and TM11 higher-order modes in the common waveguide. In this way, the operative frequency band of the device can be extended above the cutoff frequency of these modes up to the TE20 cutoff. The two symmetric side arms have to be combined using both straight and bent rectangular waveguide sections to obtain a single signal at port 4. The corresponding structure is therefore more complex than an OMT with a single side arm.

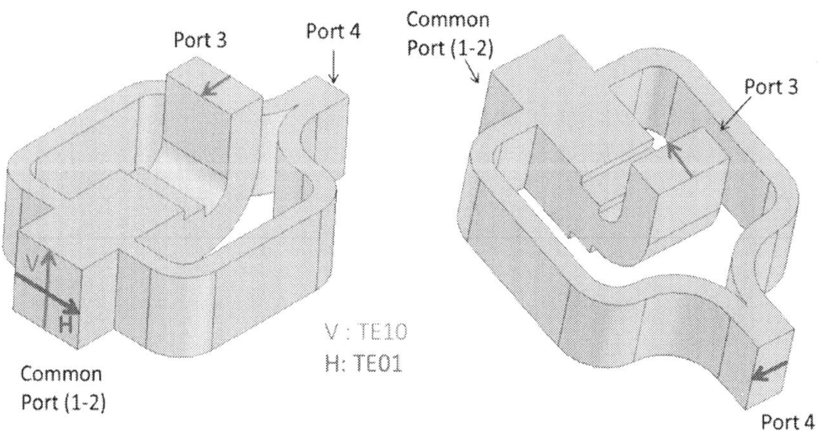

Fig. 5.10. Scheme of the Boifot OMT.

A septum (not shown) is also inserted in the common waveguide, between the coupling apertures and the stepped transition to port 3, in order to improve the matching of the H-polarization. The septum is oriented to allow the direct routing of the vertical polarization to port 3. In the original configuration, metallic posts were also inserted in the coupling apertures of the two side arms (Boifot, 1990).

It has been shown in the literature that large matching and isolation bandwidths (30%) can be obtained using this configuration. The main drawbacks consist in the manufacturing complexity and large size of the OMT. It should be pointed out that a differential error in the length of the two waveguides of the combination structure (owing to manufacturing uncertainties) can destroy the symmetry of structure with a consequent reduction of the isolation performance. Moreover, the insertion loss and group delay are intrinsically very different for the two polarizations.

5.3 Turnstile junctions

The turnstile junction (Navarrini and Plambeck, 2006) exploits a symmetric E-plane coupling for both polarizations. With reference to Fig. 5.4, the vertical polarization is only coupled to the fundamental TE10 mode at both Port 3 and Port 3'. The same polarization would also couple to the TE01 mode at ports 4 and 4'. However, this mode is under cut-off in the operative frequency range of the structure. The horizontal polarization is instead coupled to both Port 4 and 4'. It should be noted that in the E-plane coupling, the symmetric ports exhibit an opposite orientation of the electric field.

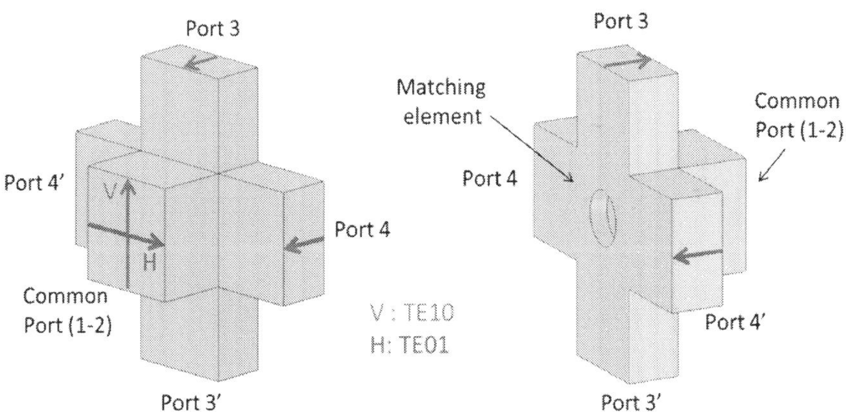

Fig. 5.11. The turnstile junction.

This turnstile junction does not excite the TE11 and TM11 modes in the common waveguide. Therefore, the upper limit of the frequency band is related to the cutoff frequency of the TE20 mode and to the cutoff frequency of the TE01 mode at the coupled ports.

A proper protrusion with either pyramidal, cylindrical or parallelepiped shape should be introduced in the back of the junction (see Fig. 5.4) in order to improve the matching.

The turnstile junction exhibits the same insertion loss and group delay for both polarizations since the latter undergo a symmetric coupling at the same section of the common port. As a drawback, two different waveguide structures (not shown) are required to combine the opposite ports. Even in this case, possible asymmetries of the combiners owing to the manufacturing uncertainties should be managed to avoid isolation problems.

This OMT type can operate in a large frequency band (more than 30%) with good power handling properties. However, the presence of two combiners make this configuration less compact and with higher losses with respect to the previous solutions.

5.4 Orthomode Junctions (OMJ)

In the case of dual-band dual-polarization feed systems where the transmit and receive bands are suitably separated, an interesting configuration is represented by the so called orthomode junction (OMJ) (Garcia, et al., 2010). Similarly to the turnstile junction, the OMJ also exploits a symmetric coupling section for both polarizations. A simplified H-plane implementation is shown in Fig. 5.5. The OMJ however exhibits a secondary common port in square (or circular) waveguide. Such a waveguide is below cut-off at the lower frequencies. Therefore, the low-band signals can be properly reflected and coupled to the side ports. Two combiners are required to obtain a single port for each polarization. It should be noted that the absence of a proper matching element in the common port leads to a quite narrow matching bandwidth for the side-coupled signals.

As far as the high-band is concerned, the complete OMT should be equipped with proper stop-band filters (not shown) on the side arms in order to prevent leakage of the high-frequency signals from the side ports. In this way, both polarizations are routed to the secondary common port. The latter can be now separated using another single-band OMT (not shown).

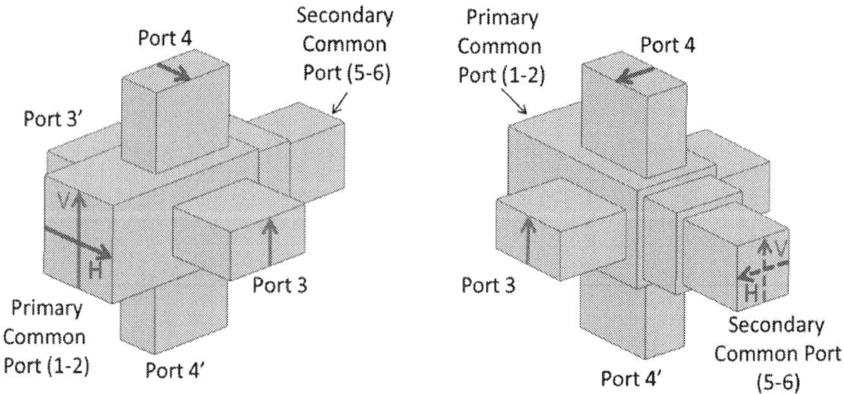

Fig. 5.12. Scheme of an Ortho-Mode Junction (OMJ).

The OMT configuration of Fig. 5.5 can be referred as a self-diplexing structure. This kind of components is very important in order to reduce the overall number of antennas on the payload. As a matter of fact, besides the narrow bandwidth, this added functionality leads to increased complexity, size and losses of the device.

5.5 Reverse coupling OMT

The broadband operative condition of some of the above-mentioned OMTs is mainly obtained inserting proper matching elements such as septa, irises or other protruding objects in the common waveguide. Besides the increased manufacturing complexity, the presence of these matching structures can limit the power handling capability of the OMT. An alternative solution to obtain broad-band OMTs has therefore been presented in (Peverini, et al., 2006). The core of the device shown in Fig. 5.6 consists in a reverse coupling section. As far as the vertical polarization is concerned, the signal in the common waveguide is coupled to the adjacent parallel rectangular waveguide by means of the E-plane apertures. This operation, which resembles the working principle of a branch-guide directional coupler, has been schematized in Fig. 5.7.

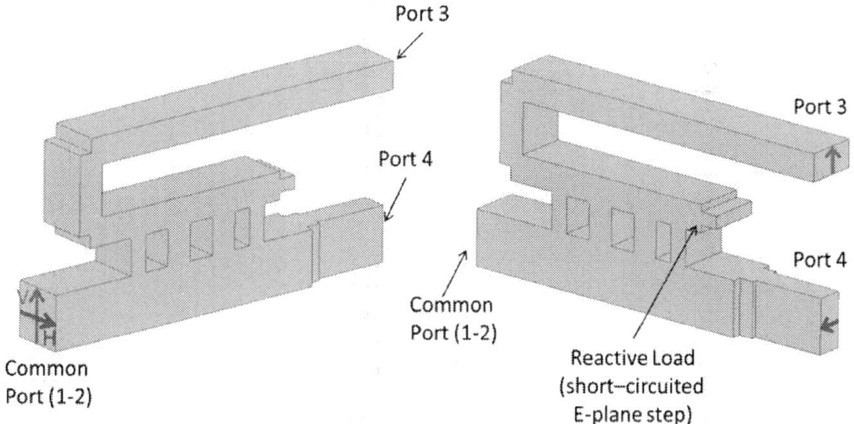

Fig. 5.13. Reverse-coupling OMT.

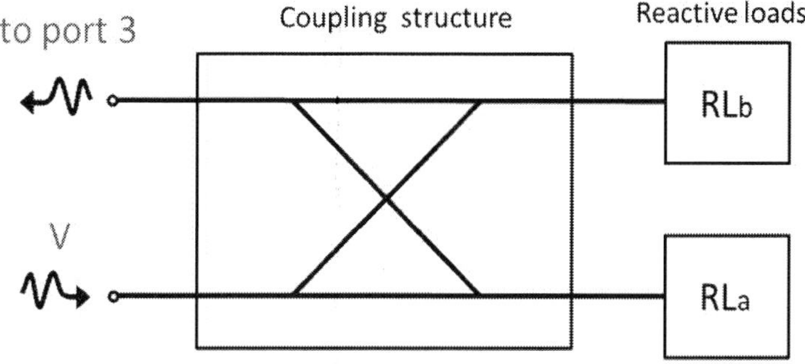

Fig. 5.14. Network representation of the reverse coupling structure for the V-polarization.

Such a directional coupler is loaded with two reactive impedances RL_a and RL_b representing the stepped transition to Port 4, which is under cut-off for the vertical polarization, and the short-circuited E-plane step on the coupled rectangular waveguide, respectively. The complete structure is properly designed so that the various coupled and reflected contributions produce a constructive interference (in-phase combination) for the V-signal to port 3. On the contrary, a destructive interference phenomenon is instead exploited to obtain a low-reflection coefficient at both common and coupled ports (Peverini, et al., 2006).

The reverse-coupling section and the stepped transition to port 4 should also be designed in order to route the horizontal polarization to port 4 with a low reflection coefficient.

The 180° bend and the subsequent straight rectangular waveguide section in Fig. 5.6 allow a proper alignment between port 3 and port 4. Furthermore, stepped waveguide twist (Baralis, et al., 2005) can be introduced to provide the same orientation of the two ports.

It should be noted that the reverse coupling structure can also be adopted to either provide a symmetric coupling structure (Navarrini and Nesti, 2009) which allow a larger operative frequency range or to design a self-diplexing unit with a more controlled broadband coupling with respect to the canonical OMJ.

6. Corrugated horns

A corrugated horn is the most employed illuminator for parabolic, offset or Cassegrain configurations in satellite feed system for its excellent potential dual polarized characteristics. The first studies on these antennas date back to the pioneer works of Clarricoats and Olver (Clarricoats and Olver, 1984). This antenna configuration originates from the theoretical study of the modes of a cylindrical waveguide where the metallic walls are substituted by a surface impedance. If specific impedance conditions are considered, the structure can support a particular hybrid mode, known as HE_{11}, whose field components, if it radiates, minimize cross polarization level. It has been shown that this particular surface condition can be realized by means of $\lambda/4$ depth corrugations. To excite this mode a suitable transition between the smooth circular waveguide and the corrugated one is necessary. This can be obtained in two ways as shown in Fig. 6.1, i.e. by means of depth corrugation increment up to the desired $\lambda/4$ value (Fig. 6.1a) or a depth corrugation decrement from the value $\lambda/2$ up to $\lambda/4$ (Fig. 6.1b). The second configuration permits wide band performances and for this reason it is usually employed. In order to satisfy the radiation pattern requirements in terms of half power beamwidth and field taper at a specific illumination angle, the radiating cross section has to be much larger than the input monomodal waveguide and therefore a suitable radius transition is necessary. The radius profile as well as the corrugations geometry are free design parameters which has to be chosen in order to match the structure and, at the same time, perform the desired conversion of the incident field to the HE_{11}-like mode. Since the number of corrugations can be of the order of hundreds, the design is quite complicate in particular for wideband application where also the antenna compactness is often required.

A part few works which gives some useful design criteria and design map (Granet et al., 2005), the standard approach in the technical literature is based on the employment of a particular radius profile as a starting point for global optimization schemes (Jamnejad et al. , 2004). In this respect, the so called dual-profile circular corrugated horn (DPCCH) is usually regarded as the state of the art. This profile consists of a combination of a sine square law and an exponential function joined by a smooth transition (see Fig.6.2). The other geometrical parameters, i.e. the dimensions and reciprocal distances of each corrugation, are usually chosen in accordance to empirical/semi analytical formulas. Although the performances obtained in this way are generally interesting, they cannot meet the specifications in the case of high performance wideband systems. For this reason global optimization algorithms (e.g. particle swarm optimization or genetic algorithms) are used not only as simple refinement tools but as a way to actually define the whole antenna geometry. The relevant drawbacks are related not only to the quite long computation times required but, mainly, to the design itself. Indeed, quite often the initial smoothness of the DPCCH profile is completely lost, which turns into a high sensitivity of the electromagnetic performances to the mechanical tolerances.

Recently a suitable design strategy has been proposed (Addamo et al. , 2010) for circular corrugated horn and here briefly described. Roughly speaking, from a functional point of view the first group of corrugations (called ``throat region'') in the horn is designed in order to convert the input incident field into the $HE_{11}$-like mode. The remaining part (called ``radiating region'') modifies this field configuration in order to guarantee the desired radiation pattern specifications (see Fig.6.3). The idea, then, is to separate the design of the throat and radiating regions by applying the most appropriate technique for each. As far as the radiating region is concerned, since the radius variation between two adjacent horn corrugations is usually relatively small, a companion periodic structure can be used (see Fig. 6.4). The desired field configuration can be then interpreted as a particular Bloch wave and the design can be obtained exploiting the periodic structure theory. The throat region definition is much more complicate since it has to perform a suitable mode conversion form the input TE_{11} to the desired HE_{11}-like mode. However since the radiating region is defined in the previous design step, this part can be obtained by means of a guided parametric analysis and therefore optimization techniques can be employed just as a refinement.

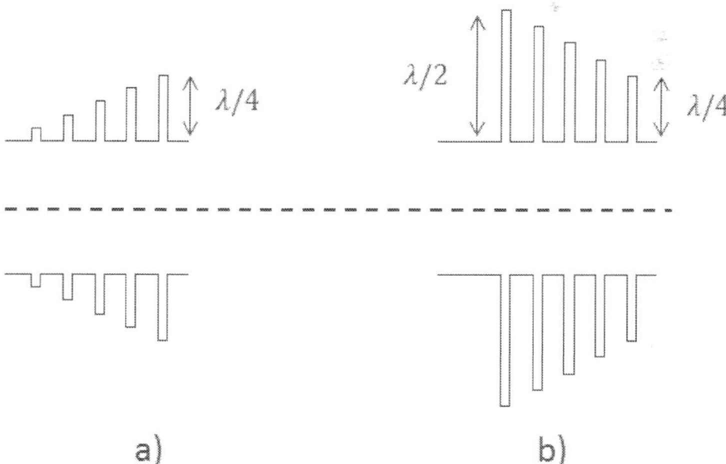

Fig. 6.15. Transitions from circular to corrugated waveguide.

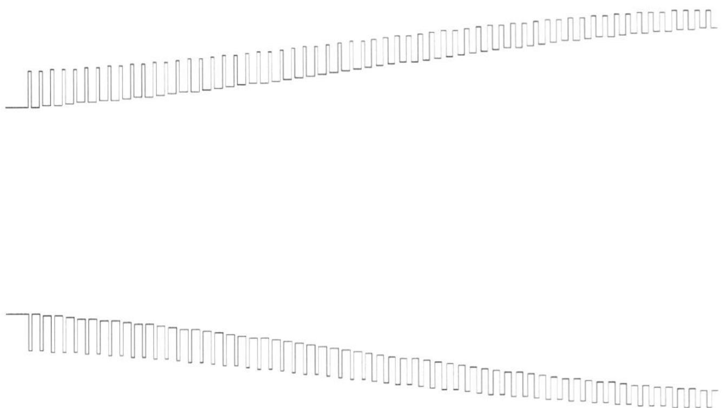

Fig. 6.2. An example of Dual Circular Corrugated Horn Profile (DPCCH).

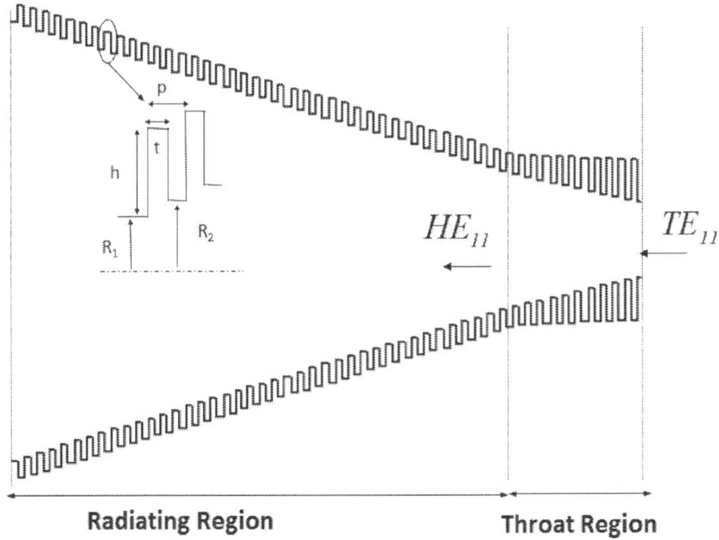

Fig. 6.3. Throat and radiating regions.

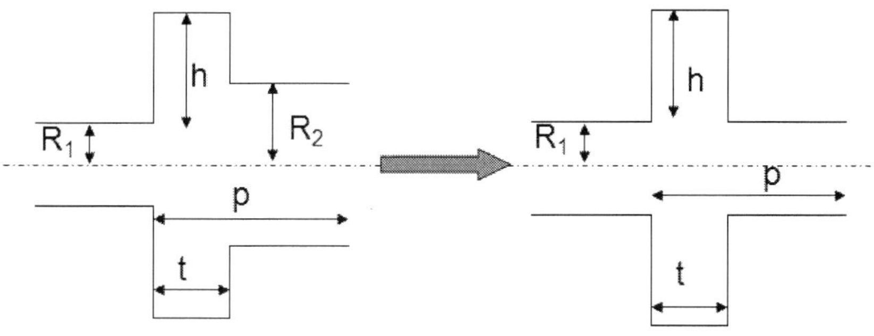

Fig. 6.4. Companion periodic structure.

7. References

Addamo G., Peverini O.A., Virone G., Tascone R., Orta R. and Cecchini P., "A Ku-K Dual-Band Compact Circular Corrugated Horn for Satellite Communications", IEEE Antennas and Wireless Propagation Letters: Volume 8, 2009, Page(s):1418 - 1421

Addamo G., Peverini O.A., Tascone R., Virone G., Cecchini P., Mizzoni R. and Orta R., "Dual use Ku/K band Corrugated Horn for Telecommunication Satellite", European Conference on Antennas and Propagation (EUCAP), Barcelona (Spain) 2010

Anza S., Vincente C., Raboso D., Gil J., Gimeno B., & Boria V. E. (2008), Enhanced Prediction of Multipaction Breakdown in Passive Waveguide Components Including Space Charge Effects, Proceedings of the 2008 IEEE International Microwave Symposium, Atlanta (U.S.), pp. 1095- 1098, June 2008

Arndt F., Beyer R., Reiter J.M., Sieverding, T., Wolf, T., "Automated design of waveguide components using hybrid mode-matching/numerical EM building-blocks in optimization-oriented CAD frameworks-state of the art and recent advances", IEEE Transactions on Microwave Theory and Techniques, Vol. 45, Issue 5, May 1997 , pp. 747-760

Baralis, M., Tascone, R., Olivieri, A., Peverini, O.A., Virone, G., Orta, R., "Full-wave design of broad-band compact waveguide step-twists", IEEE Microwave and Wireless Components Letters, Vol. 15 , Issue 2, Feb. 2005, pp. 134-136

Beniguel Y. ,Berthon A., Klooster C.V., Costes L., "Design realization and measurements of a high performance wide-band corrugated horn", IEEE Transactions on Antennas and Propagation, Volume 53, Issue 11, Page(s) 3540 - 3546, Nov. 2005

Boifot A.M., Lier E., Schaug-Pettersen T., "Simple and broadband orthomode transducer", IEEE Proceedings, Vol. 137, Pt. H, No. 6, Dec 1990, pp. 396-400

Bornemann J., Arndt F., "Transverse Resonance, Standing Wave, and Resonator Formulations of the Ridge Waveguide Eigenvalue Problem and Its Application to the Design of E-Plane Finned Waveguide Filters", IEEE Transactions On Microwave Theory And Techniques, Vol. 38, No. 8, August 1990

Cecchini P., Mizzoni R., Ravanelli R., Addamo G., Peverini O.A., Tascone R. and Virone G., "Wideband Diplexed Feed Chains for FSS + BSS Applications", EuCAP Conference 2009, Berlin (Germany), Page(s):3095 - 3099

Cecchini P., Mizzoni R., Ravanelli R., Addamo G., Peverini O. A., Tascone R., & Virone G. (2010), Ku/K Band Feed System for Satellite Applications, Proceedings of the 32nd ESA Antenna Workshop, ESTEC, Noordwijk (Netherlands), Oct.2010

Clarricoats P. J. B., Olver A.D., Corrugated Horn for Microwave Antennas, Peter Peregrinus Ltd, London (UK), 1984.

Dunning A., Srikanth S., Kerr A. R. "A Simple Orthomode Transducer for Centimer to Submillimeter Wavelengths", 20th International Symposium on Space Terahertz Technology, Charlottesville, 20-22 April 2009, pag. 191-194

European Space Agency (2007), Multipactor Calculator, Available from <http:/multipactor.esa.int/>

Garcia R., Mayol F., Montero J. M, Culebras. A. "Circular Polarization Feed with Dual Frequency OMT based on Turnstile junction", IEEE Antennas and Propagation Society International Symposium, 2010, 11-17 July 2010

Goussetis G. and Budimir D., "E-Plane Double Ridge Waveguide Filters and Diplexers for Communication Systems", European Microwave Conference, 2001, 31st Oct. 2001, Page(s):1-4

Granet C., and James G. L., "Design of corrugated horns: A primer, IEEE Antennas and Propagation Magazine, vol. 47, no. 2, pp. 76-84, April 2005.

Hartwanger C., Gehring R,, Hong U., Wolf H. and Drioli L.S., "A Dual Polarized Wide Band Feed Chain for FSS and BSS Satellite Services", EuCAP 2007 conference, Page(s)1 - 6, Nov. 2007

Jamnejad V., and Hoorfar A., "Design of corrugated horn antennas by evolutionary optimization techniques", IEEE Antennas and Wireless Propagation Letters, vol. 3, 2004, pp. 276-279.

Kirilenko A. A., Rud L. A. , Senkevich S. L. ,"Spectral Approach to the Synthesis of Bandstop Filters", IEEE Trans. Microwave Theory Tech., vol.42, no.7, Jul. 1994, pp. 1387-1392

Levy R., Cohn, S. B., "A History of Microwave Filter Research, Design, and Development", IEEE Trans. Microwave Theory Tech., vol.32, no.9, Sep. 1984, pp. 1055-1067

Levy, R. , "Compact Waveguide Bandstop Filters for Wide Stopbands", IEEE MTT-S International Microwave Symposium Digest, 2009, 7-12 June 2009, pp. 1245-1248

Lui P.L., "Passive intermodulation interference in communication systems", Electronics & Communication Engineering Journal, Vol. 2 Jun 1990, Page(s) 109-118

Navarrini A. and Plambeck R. L. , "A Turnstile Junction Waveguide Orthomode Transducer", IEEE Transactions on Microwave Theory and Techniques, Volume : 54, Issue:1 , Jan. 2006 pp. 272-277

Navarrini A., Nesti R., "Symmetric Reverse-Coupling Waveguide Orthomode Transducer for the 3-mm Band", IEEE Transaction on Microwave Theory and Techniques, Vol. 57, No. 1, Jan 2009, pp. 80-88

Parikh K. S., Singh D. K., Praveen Kumar A., Rusia S., & Sangeetha K. (2003), Multi-Carrier Multipactor Analysis of High Power Antenna Tx-Tx Diplexer for SATCOM Applications, Proceedings of the 4th International Workshop on Multipactor, Corona and Passive Intermodulation in Space RF Hardware, ESTEC, Noordwijk (Netherlands), Sept. 2003

Peverini O. A., Tascone R., Baralis M., Virone G. , Trinchero D. and Orta R., "Reduced-Order Optimized Mode-Matching CAD of Microwave Waveguide Components", IEEE Trans. Microwave Theory Tech., vol.52, no.1, Jan. 2004, pp. 311-318;

Peverini O. A. , Tascone R., Virone G., Olivieri A., Orta R., "Orthomode Transducer for Millimeter-Wave Correlation Receivers", IEEE Transactions on Microwave Theory and Techniques, Vol. 54, No. 5, May 2006, pp. 2042-2049

Peverini O.A., Tascone R., Virone G., Addamo G., Olivieri A. and Orta R., "C-Band Dual-Polarization Receiver for the Sardinia Radio-Telescope", International Conference on Electromagnetics in Advanced Applications (ICEAA09), 2009, Turin (Italy), Page(s):186 - 187;

Rebollar, J.M.; Esteban, J.; De Frutos, J.; "A dual frequency OMT in the Ku band for TT&C applications", IEEE Antennas and Propagation Society International Symposium, 1998, Vol 4, 1998 , pp. 2258 - 2261

Rozzi T. E. , "Equivalent Network for Interacting Thick Inductive Irises", IEEE Transactions on Microwave Theory and Techniques, , May 1972, Vol. 20, Issue 5, pp. 323-330

Schlegel H., Fowler W.D., "The ortho-mode transducer offers a key to polarization diversity in EW systems", Microwave System News, September 1984, pp.65-70

Tascone R., Savi P., Trinchero D., Orta R., "Scattering Matrix Approach for the Design of Microwave Filters", IEEE Trans. Microwave Theory Tech., vol.48, no.3, Mar. 2000, pp. 423-429

Tienda C., Pèrez A. M., Vicente C., Coves A., Torregrosa G., Sánchez J. F., Barco R., Gimeno B., & Boria V. E. (2006), Multipactor Analysis in Coaxial Waveguides, Proceedings of the IEEE Mediterranean Electrotechnical Conference, Benalmádena Spain, pp. 195-198, May 2006

Vahldieck R. , Bornemann J., Arndt F. ; Grauerholz D., "Optimized Waveguide E-plane Metal Insert Filters For Millimeter-wave Applications", IEEE Transactions on Microwave Theory and Techniques, Vol. 31 , Issue 1, Jan. 1983, pp. 65-69

Virone, G.; Tascone, R.; Baralis, M.; Peverini, O.A.; Olivieri, A.; Orta, R., "A novel design tool for waveguide polarizers", IEEE Trans. Microwave Theory Tech., vol.53, no.3, Part 1, Mar. 2005, pp. 888-894

Virone G., Tascone R., Baralis M., Olivieri A., Peverini O. A., Orta R., "Five-Level Waveguide Correlation Unit for Astrophysical Polarimetric Measurements", IEEE Transactions on Microwave Theory and Techniques, Volume: 55 , Issue: 2 , Part 1, 2007, pp. 309 - 317

Virone G., Tascone, R., Peverini, O.A., Addamo, G., Orta, R.,, "Combined-Phase-Shift Waveguide Polarizer", IEEE Microwave and Wireless Components Letters, 2008, Vol. 18, Issue 8, Page(s) 509 - 511

Virone G., Tascone, R., Peverini, O.A., Addamo, G., Orta, R., "Synthesis of wideband waveguide diplexers", Proceeding of the International Conference on Electromagnetics in Advanced Applications, 2009. ICEAA 2009, pp. 459 – 460

Part 7

Adaptive Antenna Arrays

7

New Antenna Array Architectures for Satellite Communications

Miguel A. Salas Natera et al.[*]
Universidad Politécnica de Madrid,
Spain

1. Introduction

Ground stations which integrate the control segment of a satellite mission have as a common feature, the use of large reflector antennas for space communication. Apart from many advantages, large dishes pose a number of impairments regarding their mechanical complexity, low flexibility, and high operation and maintenance costs. hus, reflector antennas are expensive and require the installation of a complex mechanical system to track only one satellite at the same time reducing the efficiency of the segment (Torre et al., 2006).

With the increase of new satellite launches, as well as new satellites and constellation of low earth orbit (LEO), medium earth orbit (MEO), and geostationary earth orbit (GEO), the data download capacity will be saturated for some satellite communication systems and applications. Thus, the feasibility of other antenna technologies must be evaluated to improve the performance of traditional earth stations to serve as the gateway for satellite tracking, telemetry and command (TT&C) operation, payload and payload message or data routing (Tomasic et al., 2002). One alternative is the use of antenna arrays with smaller radiating elements combined with signal processing and beamforming (Godara, 1997).

Main advantages of antenna arrays over large reflectors are the higher flexibility, lower production and maintenance cost, modularity and a more efficient use of the spectrum. Moreover, multi-mission stations can be designed to track different satellites simultaneously by dividing the array in sub-arrays with simultaneous beamforming processes. However, some issues must be considered during the design and implementation of a ground station antenna array: first of all, the architecture (geometry, number of antenna elements) and the beamforming process (optimization criteria, algorithm) must be selected according to the specifications of the system: gain requirements, interference cancellation capabilities, reference signal, complexity, etc. During implementation, deviations will appear as compared to the design due to the manufacturing process: sensor location deviation and sensor gain and phase errors (Martínez & Salas, 2010). In an antenna array, the computation of a close approach of the direction of arrival (DoA) and the correct performance of the beamformer depends on the calibration procedure implemented.

---

[*] Andrés García-Aguilar, Jonathan Mora-Cuevas, José-Manuel Fernández González, Pablo Padilla de la Torre, Javier García-Gasco Trujillo, Ramón Martínez Rodríguez-Osorio, Manuel Sierra Pérez, Leandro de Haro Ariet and Manuel Sierra Castañer.
Universidad Politécnica de Madrid, Spain

This chapter is organized with the following sections. Section 2, introduces the relationship between applications and antenna design architectures. Section 3, introduces the new antenna array architectures for satellite communication including motivation and explains experimental examples. Section 4, explains adaptive antenna array and receiver architectures for adaptive antennas systems considering the beamforming with synchronization algorithms. Finally, Section 5 explains the A3TB concept.

2. Applications and antenna design architectures

In recent effort, new antenna array architectures have been under analysis and development. In (Tomasic et al., 2002) a highly effective, multi-function, low cost spherical phased array antenna design that provides hemispherical coverage is analyzed. This kind of novel architecture design, as the geodesic dome phased array antenna (GDPAA) presented in (Tomasic et al., 2002) preserves all the advantages of spherical phased array antennas while the fabrication is based on well-developed, easily manufacturable, and affordable planar array technology (Liu et al., 2006; Tomasic, 1998). This antenna architecture consists of a number of planar phased sub-arrays arranged in an icosahedral geodesic dome configuration.

In contrast to the about 10 m diameters dome of the GDPAA, there is the geodesic dome array (GEODA) (Sierra et al., 2007) with 5 m diameters dome. This antenna, presented in Fig. 1, has two geometrical structure parts. The first one, is based on a cylinder conformed by 30 triangular planar active arrays, and the second is a half dodecahedron geodesic dome conformed by 30 triangular planar active arrays. The GEODA is specified in a first version for satellite tracking at 1.7 GHz, including multi-mission and multi-beam scenarios (Martínez & Salas, 2010). Subsequently, the system of the GEODA has been upgraded also for transmission (Arias et al., 2010).

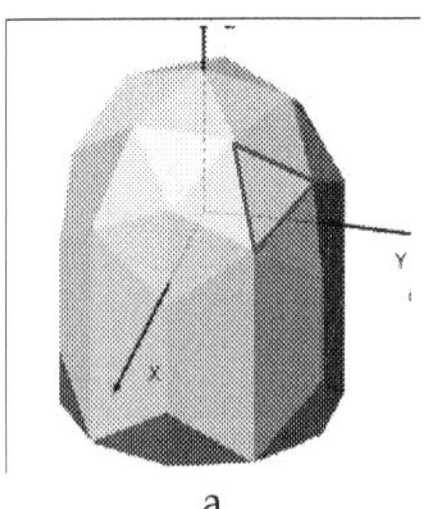

a   b   c

Fig. 1. a) The GEODA, b) The active sub-array demonstration, and c) The 45 elements planar active sub-array.

The antenna arrays technology in the user segment for satellite communications will substitute reflectors providing a more compact and easy to install antenna system, which is an interesting solution e.g. for satellite on the move (SOTM) system. There is a great diversity of solutions for fixed and mobile satellite communication systems including a large number of applications. Inmarsat broadband global area network (Inmarsat-BGAN) (Franchi et al., 2000) is the most representative example among mobile satellite systems (MSS), which gives land, maritime and aeronautical high speed voice and data services with global coverage using GEO satellites at L-band.

MSS services are divided into two groups, those that offer a regional coverage usually with GEO satellites, and those which offer a global coverage based on LEO or MEO satellite constellations. Depending on the coverage, there are some examples for MSS with regional coverage as the mobile satellite system (MSAT) in EEUU, Canada and South America, Optus in Australia, N-Star in Japan, Asia cellular satellite (ACeS) in Asia or Thuraya in the Middle East and in the North of Africa. While for MSS of global coverage there are some examples as Iridium, ICO Global Communications, Globalstar, Teledesic, etc. (Evans, 2009; Wu, 1994). Most of the MSSs work at L and S band, new applications on satellite to mobile terminal links work at X, Ku and Ka band, and satellite to base station connections work at L, S and C band. A number of applications is broad and lead terrestrial telecommunications market to offer a wider coverage: high speed voice and data (internet access, SMS, VoIP), digital video broadcasting by satellite 2 (DVB-S2) and digital video broadcasting satellite services to handhelds (DVB-SH), global position system (GPS) and Galileo, security, control and machinery monitoring on ships and aircrafts, teleeducation or telemedicine.

These modern satellite communications systems require new antenna solutions for base stations, aeronautical applications or personal communications services (PCS) on-the-move (Fujimoto & James, 2001). Within these applications, antenna array systems are potentially the best choice due to, as discussed above, its capability to perform electronically steering or beamforming, increase the antenna gain, and conform over curved or multifaceted surfaces the radiating elements. Portable antennas for PCS must be easy to install and mechanically robust, besides compact and lightweight (García et al., 2010) as the antenna array presented in Fig. 4.a. The design of antenna systems to provide high data rates for reliable PCS boarded on ships is not so strict in term of the geometrical requirements because it does not have space limitations (Geissler et al., 2010). However, in the case of land or airborne vehicles, geometrical and mechanical constraints are more severe. Antennas for terrestrial vehicles must be low profile, and for airborne vehicles aerodynamic shapes must be considered (Baggen et al., 2007; Vaccaro et al., 2010). Moreover, for the civil market conformal antenna arrays (Schippers, 2008; Kanno et al., 1996), or multi-surface arrays (Khalifa & Vaughan, 2007) are suitable choices to deal with the system aesthetic partiality.

Technological challenges have been faced during the implementation of satellite communication systems in the last decades. The design of a Test-Bed flexible and modular for testing or debugging beamforming algorithms and receiver architectures is an invaluable contribution in the educational, research and development area on satellite communication systems. The adaptive antenna array Test-Bed (A3TB) concept is based on the use of antenna arrays with beamforming capability to receive signals from LEO satellites (Salas et al., 2008). The scope of the A3TB is to probe the concept of antenna arrays applied to ground stations instead of reflectors for different applications, such as telemetry data downloading. It is also a good chance for Universities and Research Centers aiming to have their own ground station sited in their installations.

The A3TB ground station relies on the use of an antenna array to smartly combine the received signals from the satellite thanks to the implementation based on software defined radio (SDR) technology. The advantages of the SDR implementation is that A3TB architecture can be used to process any received signal from LEO satellites in the band imposed by the radio frequency (RF) circuits. Moreover, most of the processing is performed in software, so that appropriate routines can be used to process any received signal. The A3TB can be used to analyze the feasibility of different receivers and beamformer algorithms, regarding the capability to switch the receiver architecture in terms of the synchronizer algorithm configuration (Salas et al., 2007).

The current version has been developed to track The National Oceanic and Atmospheric Administration (NOAA) satellites in the very high frequency (VHF) band, in particular, the automated picture transmission (APT) channel (Salas et al., 2008). Previous versions of A3TB dealt with low rate picture transmission (LRPT) signals from the meteorological operational satellite-A (MetOp-A), where a complete receiver with beamforming and synchronization stages has been implemented (Salas et al., 2007; Martínez et al., 2007).

3. Antenna arrays for satellite communications

Satellite applications require compactness, lightweight and low cost antenna systems to be mounted on a terrestrial vehicle, an aircraft or a ship, or as a portable man-pack or a handset, and to be competitive against ground systems. Its major advantage is the possibility of getting a wider or even a global coverage. For such purposes, antenna arrays offer the technology to get a directive system whose steering direction can be electronically and/or mechanically controlled. However, planar arrays usually cannot steer more than 60°-70° from the normal direction of the antenna (Mailloux, 2005). Thus, when a wider angular coverage is required conformal arrays are an appropriate option (Josefsson & Persson, 2006). Arrays can approximate conformal shapes, such as spheres or cylinders, using several planar arrays, simplifying fabrication of active components (Sierra et al., 2007).

Since the low cost and low weight specifications are of importance, micro-strip antennas are mostly used, due to its capacity to be printed over a dielectric substrate with photolithography techniques. Low cost and low permittivity substrates are usually used such as FR4 or PTFE with different quantities of glass or ceramic impurities. For more demanding applications, ceramics, like alumina or high/low temperature co-fired ceramics (HTCC/LTTC) allow the use of smaller components thanks to its high permittivity, and give robustness against mechanical stresses and high temperatures.

3.1 Geodesic antenna array for satellite tracking in ground station

The aim of using a single antenna for tracking many satellites at the same time avoiding mechanical movements as well as its inexpensive cost make these antennas an alternative to be considered (Salas et al., 2008). Multi-beam ability and interference rejection are facilitated thanks to the electronic control system of such antennas that improves the versatility of the ground stations.

The GEODA is a conformal adaptive antenna array designed for MetOp satellite communications with specifications shown in Table 1. This antenna was conceived to receive signals in single circular polarization (Montesinos et al., 2009). Subsequently, in recent efforts the system has been upgraded also for transmission and double circular polarization (Arias et al., 2010). Hence, operating at 1.7 GHz with double circular polarization it can communicate with several LEO satellites at once in Downlink and Uplink. Current structure is the result of a comprehensive study that valued the ability to cover a given spatial range considering conformal shape surface and a given beamwidth (Montesinos et al., 2009). As Fig. 1 shows, GEODA structure consists of a hemispherical dome placed on a cylinder of 1.5 meters height. Both cylinder and dome are conformed by 30 similar triangular planar arrays (panels). Each panel consists of 15 sub-arrays of 3 elements (cells). The radiating element consists of 2 stacked circular patches with their own

RF circuits. The principal patch is fed in quadrature in 2 points separated 90° in order to obtain circular polarization. The upper coupled patch is used in the aim of improving the bandwidth.

Each panel is able to work itself as an antenna since they have a complete receiver that drives the 1.7 GHz signal to an analog to digital converter (ADC). In order to adapt the signal power to the ADC, it is mandatory to implement a complete intermediate frequency (IF) receiver consisting of heterodyne receiver with an automatic gain control block. Hence, each triangular array has active pointing direction control and leads the signal to a digital receiver through an RF conversion and filtering process. To follow the signal from the satellite, the main beam direction has to be able to sweep an angle of 60°. In this way, it is needed a phase shift in the feeding currents of the single radiating element. Previous calculations have demonstrated that 6 steps of 60 degrees are needed to achieve the required sweeping angle. An adaptive digital system allows the adequate signal combination from several triangular antennas. The control system is explained in (Salas et al., 2010).

Parameter	Specification	Parameter	Specification
Frequency range [GHz] Tx: Rx:	1.65 to 1.75 1.65 to 1.75	Isolation between Tx and Rx [dB]	>20
Polarization	Dual circular for Tx and Rx bands	VSWR	1.2:1
G/T [dB/K] For elevation >30° For elevation 5°	3 6	SLL [dB]	-11
EIRP [dBW]	36	Size [m]	1.5x1.5x3
3dB beamwidth [deg.]	5	Accuracy steering [deg.]	±1.4
Maximum gain [dBi]	29	Coverage [deg.]: Azimuth Elevation	360° >5°
Efficiency [%]	50		

Table 1. Main specifications for GEODA antenna.

3.1.1 Cell radiation pattern

Based on the study presented in (Sierra et al., 2007), the single radiating element is a double stacked circular patch that works at 1.7 GHz with 100 MHz bandwidth. In order to obtain circular polarization, the lower patch, which has 90 mm diameter, is fed by 2 coaxial cables in quadrature. Both coaxial cables connect the patch with a hybrid coupler to transmit and receive signals with both, right and left, circular polarizations. The upper patch is a circular plate with 78.8 mm diameter, and it is coupled to the lower patch increasing the bandwidth by overlapping both resonant frequencies tuning the substrate thickness and the patch diameter size. Fig. 2.a shows the radiating element scheme and main features of the layer structure are specified in (Montesinos et al., 2009).

A cell sub-array of 3 radiating elements shown in Fig. 2.b is considered the basic module to build the planar triangular arrays. The whole cell fulfills radiation requirements since it has a good polar to crosspolar ratio and a very low axial ratio. Likewise, as it is presented in Fig. 2.c, the radiation pattern shows symmetry and low side lobes for full azimuth.

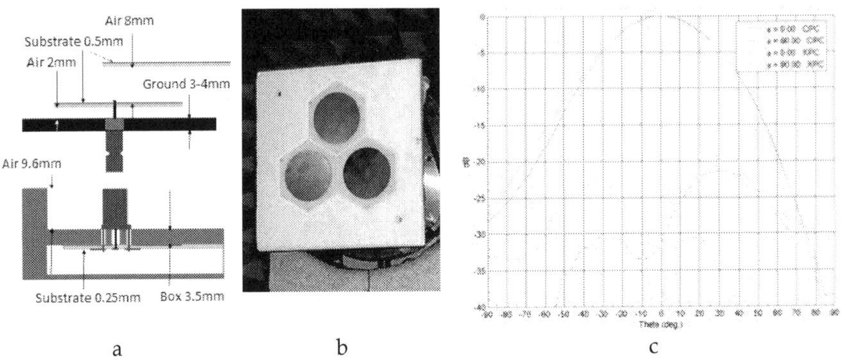

Fig. 2. a) Assembly of the single radiating element, b) Cell scheme, and c) Cell radiation pattern.

3.1.2 Transmission and Reception (T/R) module and cell distribution

Different T/R module configurations have been considered, providing either single or double polarization (Arias et al., 2010). T/R module allows amplifying and controlling the phase shift between signals, received and transmitted, providing an adaptive beam and steering direction controller in the whole working pointing range. As Fig. 3 shows, the design implemented contains a hybrid coupler, enabling double circular polarization; a double pole double throw (DPDT) switch, selecting polarization associated with transmission and reception way; 2 low noise amplifiers (LNAs), which amplify the signal received or transmitted; a single pole double throw (SPDT) switch, choosing transmission or reception way; and phase shifters, introducing multiples of 22.5° relative shift phases to form the desired beam. These surface mount devices have been chosen in order to reduce space and simplify the design.

Signals transmitted/received by the 3 T/R modules placed in a cell are divided/combined thanks to a divider/combiner circuit composed of 3 hybrid couplers that leads the signal to a general T/R module where signal is amplified. Due to transmission and reception duality, 2 SPDT switches are used to select the amplification way. Furthermore, each T/R module has associated a -25dB directional coupler that is used to test T/R modules in the transmission mode. Additionally, reception mode is tested by measuring signal in the divider/combiner circuit. A single pole 6 throw (SP6T) switch selects the path that is tested.

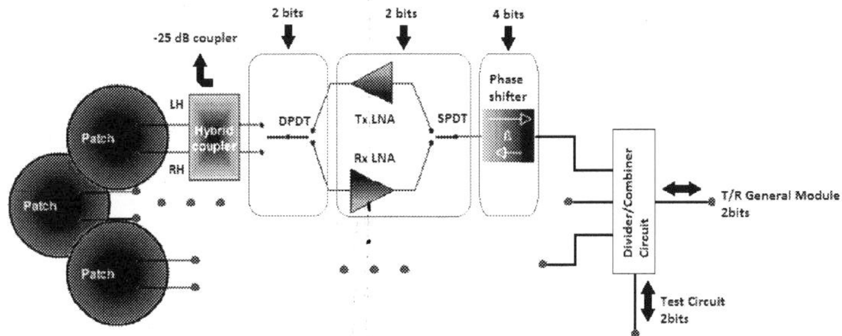

Fig. 3. Cell sub-array and RF circuit.

3.1.3 Control system

The control system has two main parts (Salas et al., 2010), the hardware structure and the control software. The two level hardware structure has the lowest possible number of elements, making the control simpler in contrast to the previous in (Salas et al., 2010). Finally, an inter-integrated circuit (I2C) expander is used to govern T/R modules individually, and one more cover cell needs (LNA of call and test). A multipoint serial standard RS-485 is used to connect the computer with the panels.

3.2 Portable antenna for personal satellite services

New fix and mobile satellite systems (Evans, 2000) require antenna systems which can be portable, low profile and low weight. Planar antennas are perfect candidates to fulfill these specifications. Usually slots (Sierra-Castañer et al., 2005) and printed elements (García et al., 2010) are most used as radiating elements.

3.2.1 Antenna system structure

In this subsection it is introduced a printed antenna for personal satellite communications at X band, in Fig. 4. Table 2 shows main antenna characteristics.

Parameter	Specification	Parameter	Specification
Frequency range[GHz] Tx: Rx:	7.9 to 8.4 7.25 to 7.75	Efficiency [%]	50
Polarization	Dual circular polarization for Tx and Rx bands	Isolation between Tx and Rx [dB]	>17
G/T [dB/K]	7	VSWR	1.4:1
EIRP [dBW]	32	SLL [dB]	-11
3dB beamwidth [deg.]	5	Size [m]	40x40x2.5
Maximum gain [dBi]	25	Weight [Kg]	2

Table 2. Portable antenna specifications.

This is a planar, compact, modular, low loss and dual circular polarized antenna, for Tx and Rx bands, simultaneously. It is made up by a square planar array of 16x16 double stacked micro-strip patches, fed by two coaxial probes. A hybrid circuit allows the dual circular polarization (Garg et al., 2001). Elements are divided in 16 sub-arrays excited by a global power distribution network of very low losses, minimizing the losses due to the feeding network and maximizing the antenna efficiency. In order to reduce side lobe levels (SLL), the signal distribution decreases from the centre to the antenna edges, keeping symmetry with respect to the main antenna axes. The antenna works at X band from 7.25 up to 8.4 GHz with a 14.7% relative bandwidth for a 1.4:1 VSWR and a maximum gain of 25 dBi.

3.2.2 Sub-array configuration

The sub-array configuration can be seen in Fig. 4.a. It makes possible to separate the fabrication of these sub-arrays from the global distribution network, simplifying the corporative network and getting a modular structure suitable for a serial fabrication process. Each sub-array is a unique multilayer board, where PTFE-Glass substrate of very low losses has been used as base material. The power distribution network is connected to each sub-array through (SMP-type) coaxial connectors.

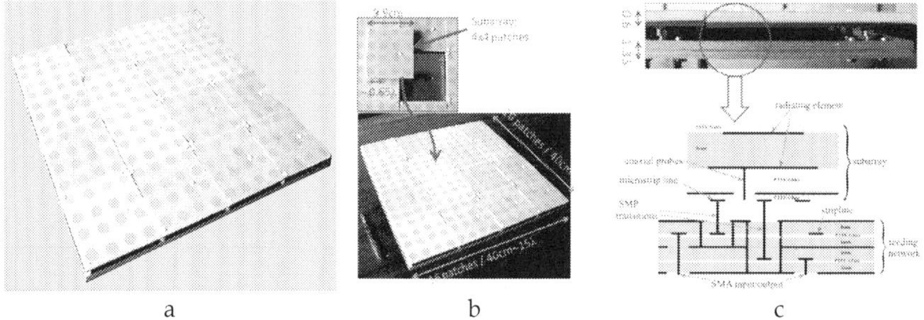

a   b   c

Fig. 4. a) Dual polarized portable printed antenna for satellite communication at X band, b) Sub-array perspective view, and c) Side view and multilayer scheme.

Fig. 5.a and Fig. 5.b show the sub-array unit cell. In order to obtain better polarization purity, each element is rotated 90° and excited by a 90° phase-shifted signal. Moreover, in Fig. 5.c is showed a miniaturized branch-line coupler (BLC) of three branches working as a wide band hybrid circuit (García et al., 2010; Tang & Chen, 2007).

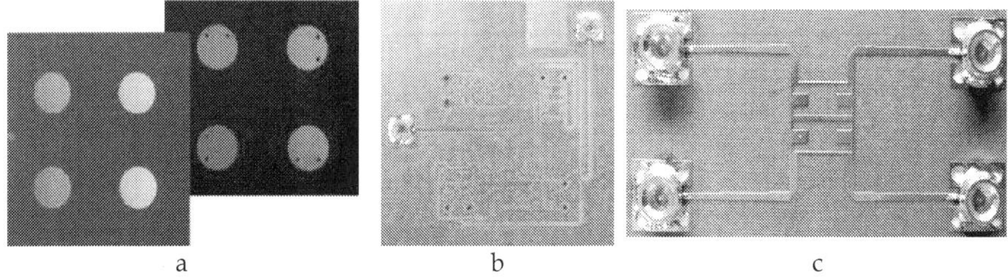

a   b   c

Fig. 5. Unit cell test board, a) Unit cell test board 2x2 stacked patches, b) Micro-strip feeding network, and c) Miniaturized BLC Prototype.

A conventional configuration takes up an area of 13.3 cm² which is big compared to the radiating element and the sub-array subsystem size. Therefore, a miniaturization of the BLC is needed using the equivalence between a $\lambda/4$ transmission line and a line with an open-ended shunt stub. An area reduction about 35% is achieved and the hybrid circuit behaves like a conventional BLC. In Fig. 6.b and Fig. 6.c measurement results for the BLC in Fig. 5.c are shown compared with simulations.

Fig. 7 depicts some sub-array measurements. The copular to crosspolar ratio is better than 25 dB and axial ratio is under 0.9 dB in the whole bandwidth.

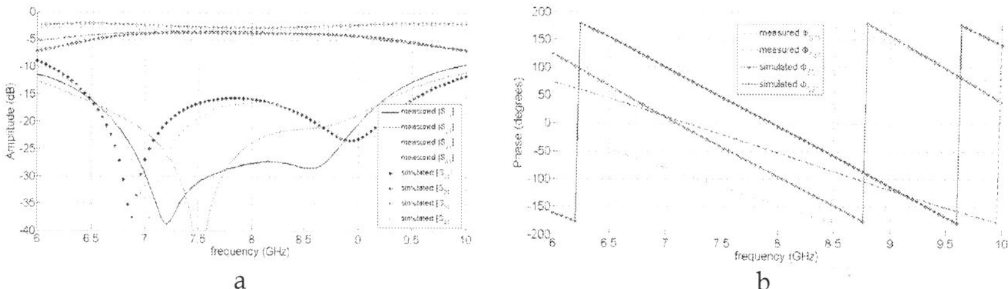

a   b

Fig. 6. Miniaturized BLC, Measured and simulated S-parameters in: a) Amplitude, and b) Phase.

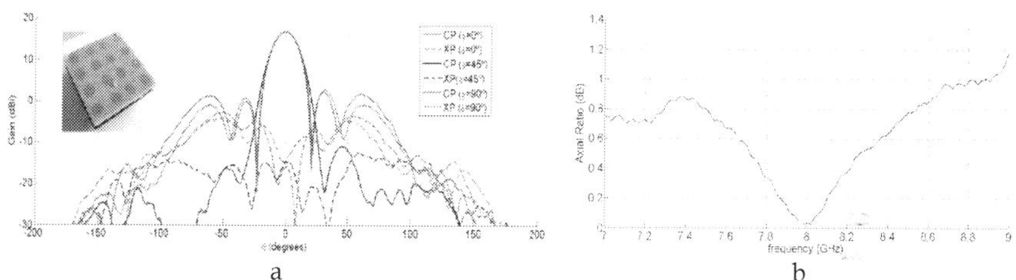

a   b

Fig. 7. 4x4 patch sub-array measurements, a) Radiation pattern at 7.75 GHz, and c) Axial ratio for right-handed circular polarization.

3.2.3 Low losses power distribution network

The global feeding network presented in Fig. 8.a is a protected strip-line, where foam sheets of high thickness are used to get low losses. Such a kind of feeding network allows keeping a trade-off between the simplicity of exciting the radiating elements using printed circuits and the loss reduction when the distribution network is separated in a designed structure to have low losses. Losses in the structure are around 0.6 dB/m which yields to 0.3 dB of losses in the line. Two global inputs/outputs using SMA-type connectors, one for each polarization, excite the strip-line networks.

Vertical transitions have to be treated carefully and must be protected to avoid undesired higher order mode excitation. Thereby, it has been design a short-ended pseudo-waveguide, adding some extra losses about 0.3 dB, for two kinds of vertical transitions, as can be seen in Fig. 8.b and Fig. 8.c.

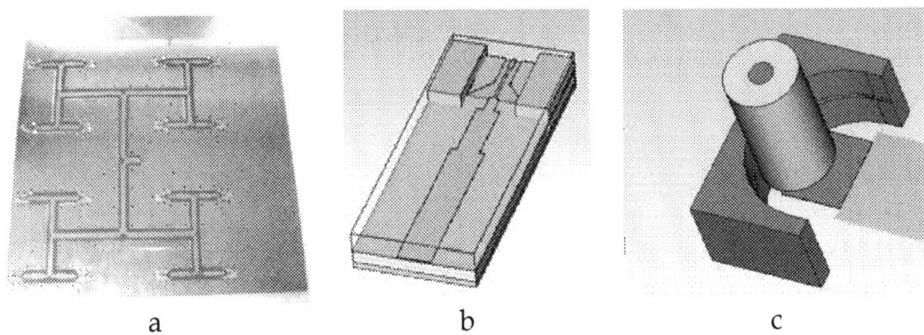

a   b   c

Fig. 8. a) Protected strip-line global corporative network for one polarization, b) Transitions from strip-line to SMA-type connector, and c) Transitions from strip-line to SMP-type connector.

3.2.4 Antenna performance

Fig. 9 depicts measured radiation pattern at 7.75 GHz, gain and axial ratio for the antenna system. It is shown a maximum gain of 25 dBi in the lower band and about 22 dBi in the upper band, and a SLL around 11 dB. Copolar to crosspolar ratio is better than 30 dB and axial ratio is under 0.7 dB. Total losses are about 4 dB in the working band.

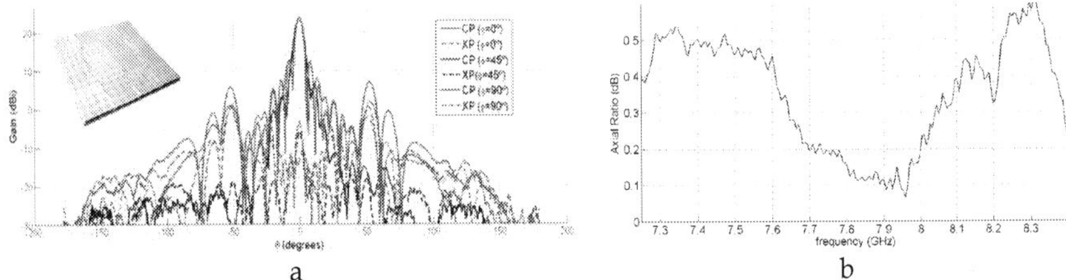

a   b

Fig. 9. Antenna measurements results, a) Radiation pattern at 7.75 GHz, and c) Axial ratio for right-handed circular polarization.

3.3 Electronically steerable antennas for mobile and fixed portable systems

At present, two types of electric steerable antenna systems can be used to access the satellite communication services (Bialkwoski et al., 1996). These are: fixed position portable systems and mobile systems such as those installed on a land vehicle. The fixed portable antenna system is relatively easy to be accomplished by the antenna designer. The design involves standard procedures that concern the operational bandwidth, polarization and moderate gain (García et al., 2010). One drawback of the fixed position portable system is that they require the user to be stationary with respect to the ground. This inconvenience can be overcome with the mobile antenna system. A mobile user complicates the scenario since the ground mobile antenna needs to track the satellite (Alonso et al., 1996). The design of such a system is more challenging as new features associated with the mobility of the system have to be incorporated (Fernández et al., 2009). The requirement leads to a narrow beamwidth, for which satellite tracking is required as the vehicle moves around. Electronically steerable antennas enable the development of reconfigurable antennas for satellite applications.

3.3.1 Steerable antenna for fixed position portable systems

This antenna is a fixed satellite communication system with high gain at X band, consisting of an antenna array that integrates 32 2x2 sub-array modules in the complete antenna, as shown in Fig. 10.a. It is a planar and dual circular polarized antenna for Tx and Rx bands simultaneously. It is made up by a planar array of double stacked circular micro-strip patches, fed by 2 coaxial probes to generate circular polarization. A hybrid circuit allows the dual circular polarization as shown in Fig. 10.b.

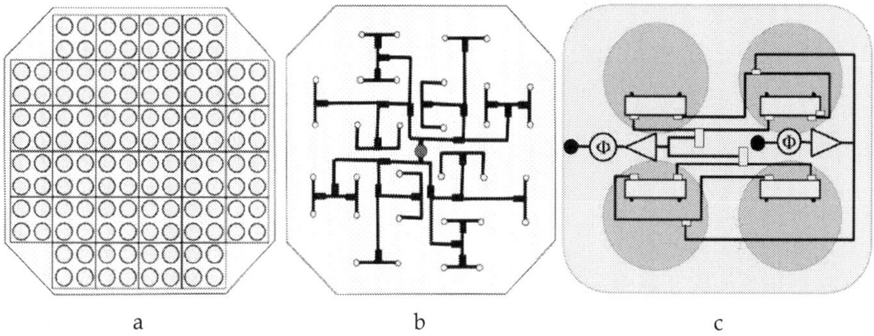

a   b   c

Fig. 10. Active multi-beam antenna, a) Top view, b) Feeding network of the complete antenna, and c) Beamforming network of the 2x2 sub-array module

The antenna has the same design parameters, structure and configuration as the antenna explained in Section 3.2 but with a different feeding network, as previously shown. In this case, the beamforming network requires changes in the feeding phase in the 2x2 sub-arrays, which can be achieved by phase shifters (ϕ) associated with different sub-arrays (Fig. 10.c). All these sub-arrays are connected to a feeding network, in Fig. 10.b, formed by transmission lines with low losses in strip-line. General specifications of the steerable antenna for fixed position portable systems are provided in Table 3.(a).

3.3.2 Automatic steerable antenna for mobile systems

A broadband circularly polarized antenna for satellite communication in X band is presented in Fig. 11 and specified in Table 3.(b). The arrangement features and compactness are required for highly integrated antenna arrays. It is desired to get a low-gain antenna for mobile satellite communications with low speed of transmission. In this system, the antennas are formed by 5 planar 4x4 arrays of antennas, which form a truncated pyramid with a pointing capability in a wide angular range, so that among the 5 planar arrays the complete antenna can cover any of the relative positions between the mobile system and the satellite in a practical way. The scheme of the active antenna can be seen in Fig. 11.

As it can be observed in Fig. 11.a, the antenna terminal is a multi-beam printed antenna shaped as a trunk pyramid capable of directing a main beam in the direction of the satellite. The antenna steering system consists of a multi-beam feeding structure with switches that lets combine the feed of each 4x4 arrays to form multiple beams. Switching the different 4x4 arrays, it is achieved different multiple beams and the variation of the steering direction.

The complete antenna consists of a Tx and Rx module that works independently in the 2 frequency bands.

The antenna has multiple beams covering the entire space to capture the satellite signal without moving the antenna. The signal detected in each of the beams is connected to a switch, which, by comparison, is chosen the most appropriate 4x4 array. The steering direction of the 4x4 array can vary between a range of directions that covers a cone angle range of 90°. To obtain the required gain and cover the indicated range, it is required around 15 beams, which can be obtained by integrating the beamforming networks with switches in the design as presented in (Fernández et al., 2009).

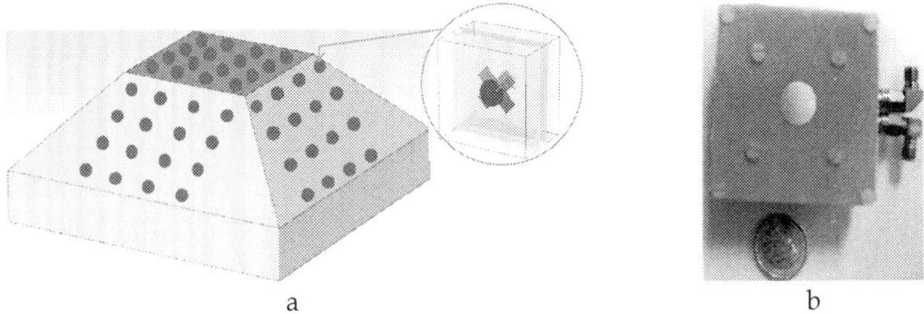

a   b

Fig. 11. Complete antenna structure, a) Radiating element of the 4x4 arrays, and b) Prototype top view.

The radiating element of the 4x4 array is one 2 crossed dipoles with a stacked circular patch as shown in Fig. 11.a and Fig. 11.b. In Fig. 12 the cross-section of the radiating element structure is presented.

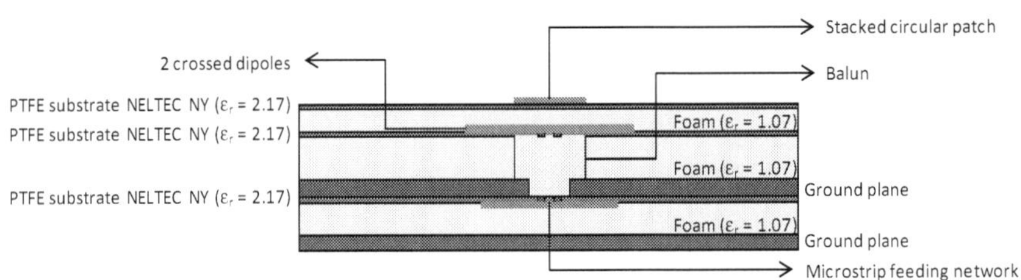

Fig. 12. Cross-section scheme of the radiating element.

The key element of the radiating element feeding structure (Fig. 14.b) is a resonant micro-strip feed ring that has been implemented, as well as a micro-strip 90° branch-line coupler to obtain the desired right hand or left hand circular polarizations (RHCP or LHCP) which ensures adequate port coupling isolation. The S-parameters in amplitude and phase of the micro-strip feeding structure are shown in Fig. 13.a and Fig. 13.b.

Fig. 14.a depicts the S-parameters of the radiating element with the micro-strip feed structure and they fulfill the specification, in Table 3.(b). In Fig. 14.c, the radiation pattern of the radiating element at 7.825 GHz is shown and in Fig. 14.d the radiation pattern of the 4x4 arrays is presented. It is shown a maximum gain of 19.4 dBi at the center frequency band (7.825 GHz). Copolar (CP) to crosspolar (XP) ratio is better than 17 dB and the axial ratio is under -3dB.

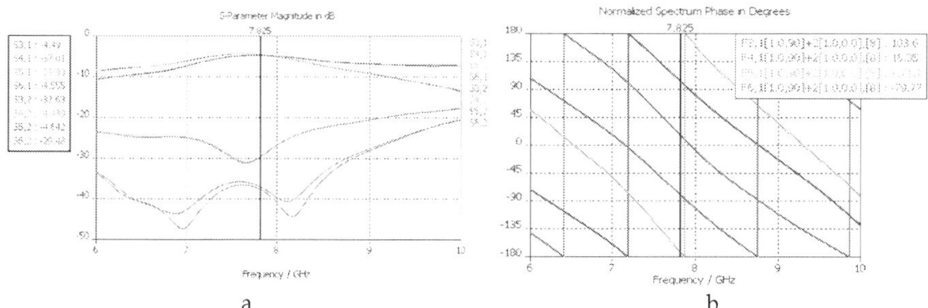

Fig. 13. Micro-strip feeding structure, a) Amplitude of S-parameters, and b) Phase of S-parameters.

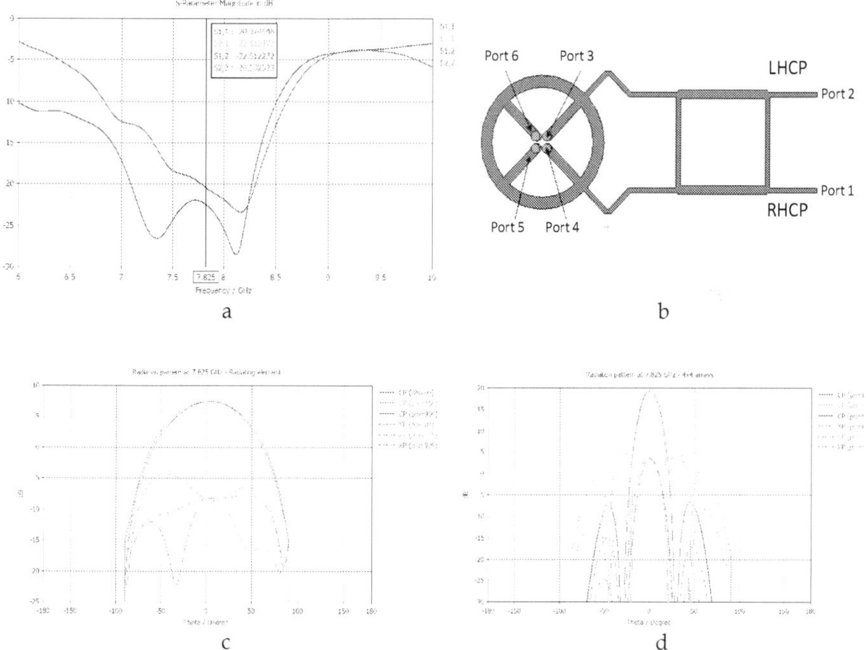

Fig. 14. a) S-parameters, b) Resonant ring + 90° branch-line coupler, c) radiation pattern at 7.825 GHz, and d) 4x4 array radiation pattern.

Parameter	Value (a)	Value (b)	Comments
Freq. range [GHz] Rx Tx	7.25 - 7.75 7.9 - 8.4	7.25 - 7.75 7.9 - 8.4	Microwave applications.
G/T (in Rx) [dB/K]	7	7	
EIRP (in Tx) [dBW]	32	32	
Beamwidth at -3dB [deg.]	4	20	
Polarization	circular	circular	In both, reception and transmission.
Gain [dBi]	>28	>15	
Axial ratio [dB]	< 1	<3	(a) Between ±50°. (b) Between ±45°.
VSWR	< 1.4:1 (-15.6 dB)	< 1.5:1 (-13.9 dB)	
Isolation between ports [dB]	< -17	< -15	
Radiation pattern [deg.]	±35	±90	Steering direction tilt.
Dimensions [cm]	40x40x4	20x20x15	

Table 3. (a) General specifications of the steerable antenna for fixed position portable systems , and (b) General features of the automatic steerable antenna for mobile systems.

3.4 Transmit-array-type lens antenna for terrestrial and on board receivers

Technology in satellite communications has revealed an increasing interest in novel smart antenna designs. Phased-array based designs are basic in electronically reconfigurable devices for satellite applications, which are more and more demanding. The strict requirements in terms of architecture, shape and robustness are important constraints for the development of planar lens-type devices. Regarding the usage and location, lens-type devices are useful for either terrestrial or on board receivers, in vehicular technology. Some clear examples are satellite communications for aircrafts preserving the fuselage aerodynamics or for some other kind of vehicles such as trains, etc.

3.4.1 Introduction to lens-type structures

In a general view, in lens-type a particular signal is received (in our case, an electromagnetic wave with specific features in terms of frequency, wave-front, etc.), it is processed (either complex signal processing techniques or only phase correction tasks can be considered in this interface), and finally, the processed signal is retransmitted.

Regarding the lens configuration, a transmit-array lens consists of three well distinguished interfaces: the first one for signal reception, one interface for signal processing, and the last one for processed signal re-radiation, as depicted in Fig. 15.

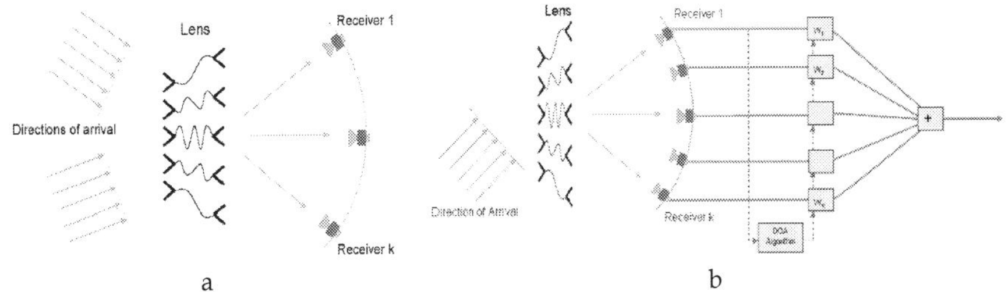

Fig. 15. a) Multi-user scheme with different receivers and transmitters, and b) Adaptive scheme with DoA determination.

These structures are intimately related to reflect-array ones, where the reception and transmission interfaces are turned to be the same interface, with a reflection-type behavior (Encinar & Zornoza, 2001). Although in an equal output phase configuration a transmit-array device behavior would be similar to the one obtained with a reflect-array, the transmit-array offers the advantage of removing the feed blockage.

In a transmission scheme, depending on the transmitter position regarding the lens, a different steering direction is achieved and a different user is pointed. In the case of reception, the situation is the same: the user position configures the direction of arrival, which determines the receiver position around the lens (Padilla et al., 2010a). In adaptive schemes, applying the proper processing algorithm to the signal received in the different receivers around the lens, it is possible to develop an adaptive steering vector, in terms of the desired direction of arrival.

3.4.2 Transmit-array lens architecture and design

Lens-type structures provide two fundamental advantages. First, phase error correction due to spherical wave front coming from the feeding antenna. Fig. 16.a shows this effect. Second, new radiation patterns configuration. Fig. 16.b depicts this fact.

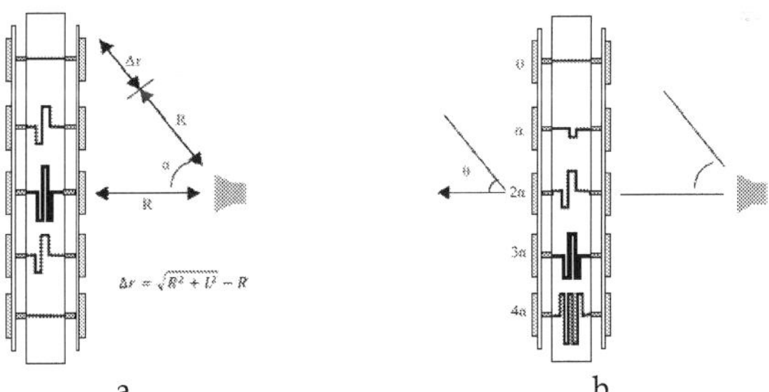

Fig. 16. a) Phase error correction, and b) Radiation pattern reconfiguration.

3.4.3 Electronically reconfigurable devices for active transmit-array lenses

The addition of reconfigurability on transmit-array devices requires the possibility of controlling the phase response of the transmitted signal at each cell of the lens. Electronic control of phase signal may be added in two different ways: First, electronic tuning of the radiating element phase response (Padilla et al., 2010a): Modifications in the radiating element circuital behavior lead to changes in phase response (arg[S_{21}]). Fig. 17 shows an electronically reconfigurable microwave patch antenna for this purpose, along with the equivalent circuit and prototype outcomes in terms of phase.

Second, electronic tuning of phase shifters in transmission lines (Padilla et al., 2010c): Modifications in the phase response of the phase shifters lead to corresponding changes in phase response. Some options are applied for these devices, such as hybrid couplers, etc. Fig. 18 shows a microwave phase shifter prototype for this purpose, along with the working scheme and its outcomes in phase.

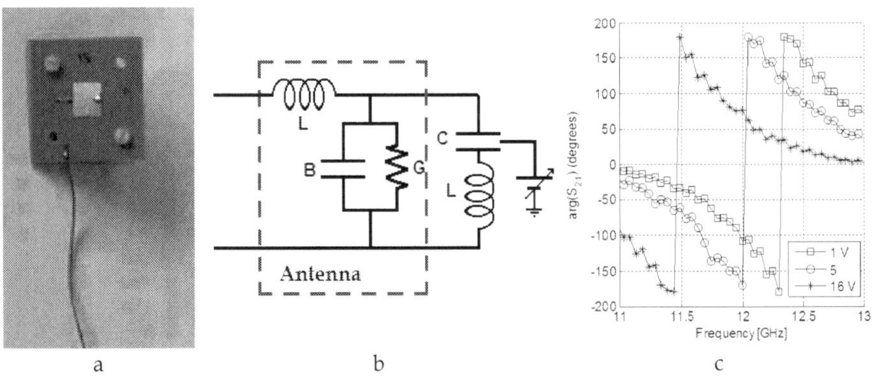

Fig. 17. Electronically reconfigurable antenna, a) Patch antenna prototypes, b) Equivalent circuit, and c) Phase behavior in frequency.

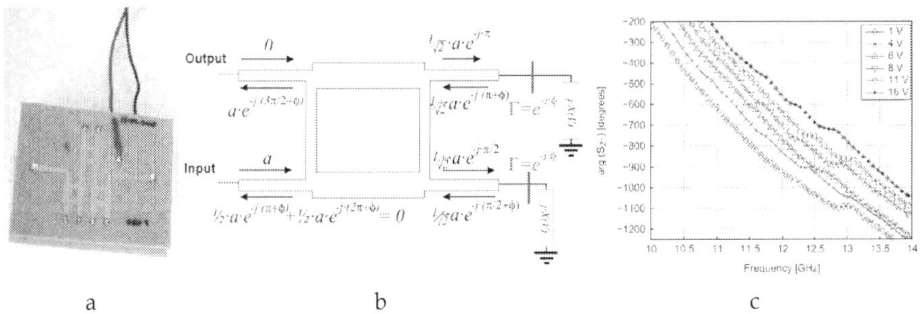

Fig. 18. Electronically reconfigurable phase shifter, a) Phase shifter prototype, b) Working scheme, and c) Phase behavior in frequency.

3.4.4 Electronically reconfigurable active transmit-array prototype

One electronically reconfigurable prototype is presented in Fig. 19 and detailed in this section. The prototype design implies the use of microwave phase shifters according to the design specified in section 3.4.3. This transmit-array lens prototype operates at 12 GHz. Main specifications are provided in Table 4.

Parameter	Value	Comments
Frequency range [GHz]	12 ± 0.5	Microwave applications.
Polarization	Linear	In both, reception and transmission.
Directivity [dBi]	>21	
Axial ratio [dB]	< 1	Between $\pm 50°$ elevation.
S_{11} [dB]	< -20	
Radiation pattern [deg.]	± 30	Steering direction tilt, for both H and V planes.
Feeding antenna [mm]	120	Corrugated horn linearly polarized
Phase shifters [deg.]	360	Full phase range variation.
Transmit-array elements	36	6x6 array topology.
Separation between elements	$0.7\lambda_0$	Related to the wavelength

Table 4. Main features of the electronically reconfigurable transmit-array prototype.

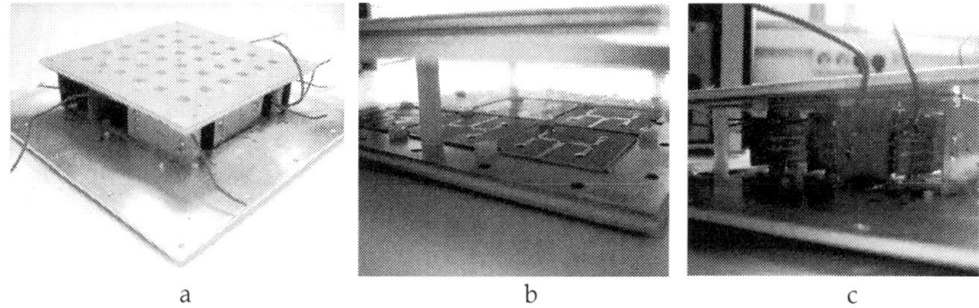

a   b   c

Fig. 19. Transmit-array core, a) Transmit-array prototype, b) Distribution networks, and c) Phase shifter integration.

The electronically controllable steering capabilities are tested and assured for a range of ± 30°in each main axis. An example of radiation pattern is provided in Fig. 20, for 9° tilt in one of the main axes.

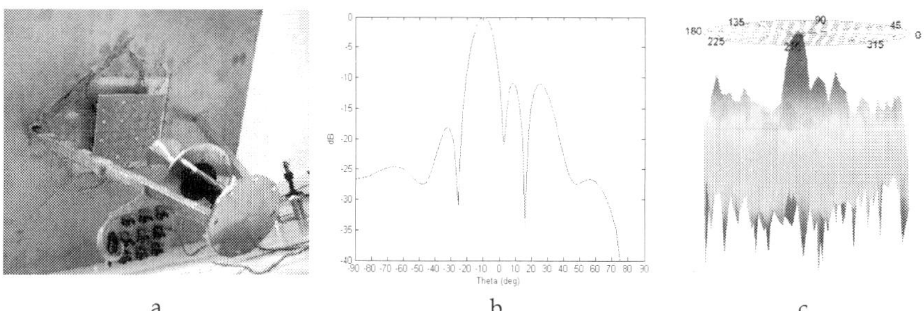

a   b   c

Fig. 20. a) Complete transmit-array with feeder and control circuits; and transmit-array measurement results for 9° tilt in one axis, b) H plane, and c) 3D plot.

4. Adaptive antenna array

Adaptive antennas can be described as systems usually based on three main parts: the antenna array, the receiver architecture and the beamforming scheme. Thus, adaptive antennas have those advantages owing to those three main parts. The system capabilities increase as complexity and development cost do. Furthermore, since signal processing is the basement of the adaptive antenna concept it is important to analyze the design challenges in terms of hardware architecture and components such as processors and embedded systems.

The antenna array provides the capability of performing the antenna pattern meeting the environment requirement under study. Besides, receiver architectures have some interesting advantages depending on the implemented receiver arraying technique such as signal to noise ratio (SNR) and bit error rate (BER) performance enhancement. Furthermore, symbol synchronization and carrier recovery can be used increasing the receiver complexity but providing higher performances. Finally, beamforming schemes use multiple antennas in order to maximize the strength of the signals being sent and received while eliminating, or at least reducing, interference as discussed in Section 4.3.

Adaptive antenna arrays are often called Smart Antennas because they have some key benefits over traditional antennas, by adjusting traffic patterns, space diversity or using multiple access techniques. The main four key benefits are: First, enhanced coverage through range extension by increasing the gain and steering capability of the ground station antenna; Second, enhanced signal quality through multi-target capability and reduction of interferences; finally, adaptive antennas improve the data download capacity in the ground segment of satellite communication by increasing the coverage range (Martínez et al., 2007).

4.1 Design and architecture based on software defined radio

For design there is the well known waterfall life cyclic model (Royce, 1970) that can be used to manage main aspects of the design of architectures. Thus, some tasks must be fulfilled subsequently as follow in Fig. 21.a.

Fig. 21.b shows the design schemes resulting of the requirement analysis stage corresponding software and hardware system specifications. In the depicted scheme, there are some system components such as the radiating element and RF circuits that are often designed under iterative prototyping model.

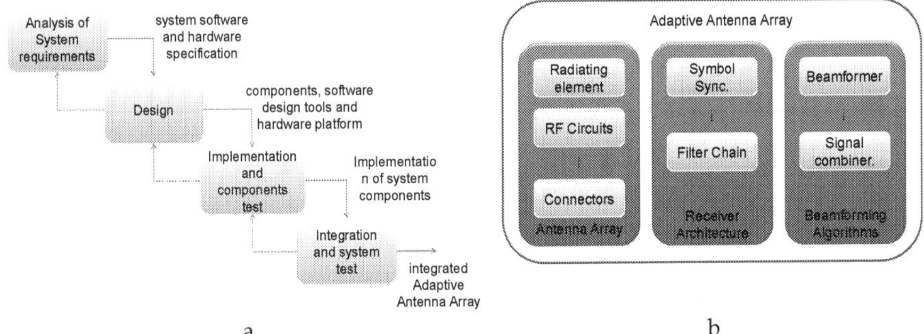

Fig. 21. a) Water life cyclic model of the adaptive antenna array design, and b) Simplified design scheme of adaptive antenna arrays.

Regarding the hardware implementation, tables presented in (Martínez et al., 2007) show the hardware resource consumption in the field programmable gate array (FPGA) Virtex-4 for the least mean squared (LMS) beamforming algorithm with full spectrum combining (FSC) receiver architecture and SIMPLE beamforming algorithm with symbol combining (SC) receiver architecture. Both scheme designs have an antenna array of 2 elements. The algorithm based on correlation requires less hardware. The main difference can be appreciated in the amount of digital signal processing oriented component (DSP48) resources, typically used for filtering applications (Martínez et al., 2007).

4.2 Receiver architectures based on algorithms type

Several receiver architectures can be implemented, and they are frequently based on the type of the beamforming algorithm used. When training signals are available in the transmitted frame, a time-based reference algorithm can be used. However, this solution is only valid when the earth station is capable of demodulating the received training sequence. Other algorithms used in deep space communications are based on signal correlation and they avoid performing the demodulating process. This kind of algorithms are blind techniques that do not require any additional signal demodulation before applying some beamforming technique and work better in low SNR conditions than time-based algorithms. Several receiver architectures can be implemented exploiting the processing capabilities of the SDR, such as FPGA, application-specific integrated circuits (ASICS), and digital signal processing (DSPs). The design of the receiver architecture fundamentally depends on the selection of beamforming algorithms. An example of beamforming technique is the LMS algorithm whose estimation of coefficients or weights requires a temporal reference and is implemented through SC receiver architecture (Fig. 22.a). In the other hand, the SIMPLE algorithm (Rogstad, 1997) constitutes a beamforming technique that is implemented using FSC receiver architecture (Fig. 22.b) in order to perform the calculation of weights.

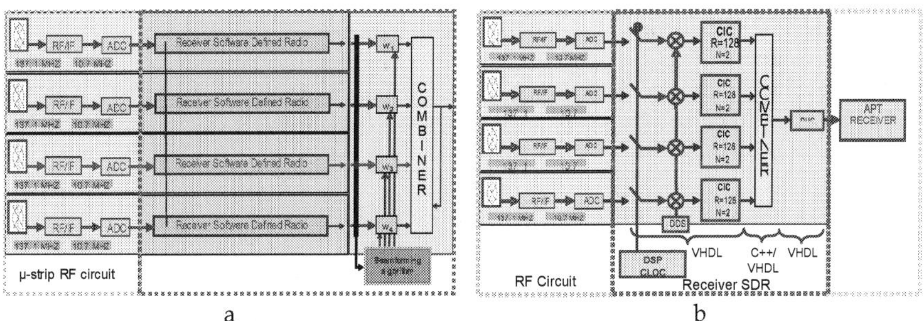

a   b

Fig. 22. Comparison of receiver architectures. a) Symbol Combining (SC), and b) Full Spectrum Combining (FSC).

The SC architecture can be divided into two more sub-classes which work on a phase-recovery basis. The complex symbol combining (CSC) recovers the phase information with regard to a reference element using feed-forward and feedback algorithms. One of the advantages of this scheme is that the rate of data sent to the combining module has a rate slightly higher than the symbol rate. For most applications, the symbol rate is relatively low and is a multiple of the data rate. In this kind of schemes, there is an important cost consideration in real-time applications and the requirements of instrumental phase stability are very severe (Rogstad et al., 2003). Other type of SC architecture is the stream symbol combining (SSC). In this kind of scheme, data are sent to the combining module at a rate equal to the symbol rate. The symbol rate depends on the coding scheme and for most applications is relatively modest. Also, the requirements of instrumental phase stability are no severe, as in the case of CSC scheme. The disadvantage of the SSC is the additional hardware required for each antenna.

Furthermore, there are the baseband combining (BC) and carrier arraying (CA) architectures discussed in (Rogstad et al., 2003). In BC architectures the signal from each antenna is carrier locked and combining in baseband for further demodulation and synchronization. In effect, the carrier signal from the spacecraft is used as a phase reference so that locking to the carrier eliminates the radio-frequency phase differences between antennas imposed by the propagation medium. Besides, in CA architectures, one individual carrier-tracking loop is implemented on each array element. Then, the elements branches are coupled in order to increase the carrier-to-noise ratio (CNR), but losses of radio channel are far compensated (Rogstad et al., 2003).

In general, the selection of the beamforming algorithms is determined by the following aspects: Hardware and computational resources; Speed of convergence and residual error of adaptive algorithms; Calibration requirements and auto-compensation ability; and system signal-transmission characteristics.

4.3 Beamforming techniques for satellite tracking

Some satellites transmit useful information inside its frames for synchronization and tracking purposes. The gathering of satellite data requires the tracking operation along its earth orbit. To accomplish this goal with adaptive array architectures, some beamforming techniques should be implemented. Fig. 23 illustrates a simple example of a narrowband linear adaptive beamformer system.

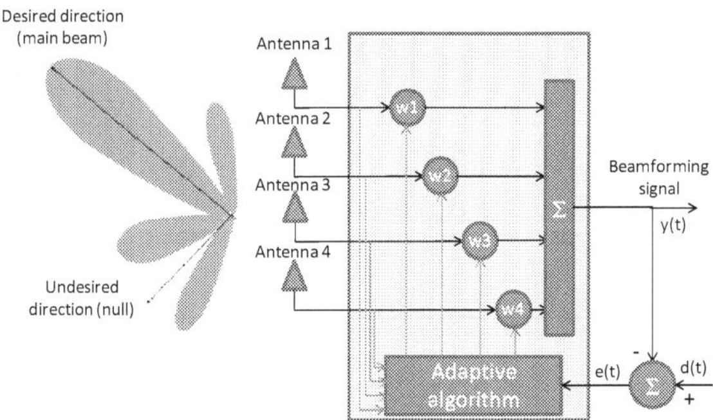

Fig. 23. Adaptive antenna system.

A linear beamformer combines signals according to some weights w_i, to produce a desired radiation pattern. The mathematical expression of a linear beamformer at the array output in vector notation can be expressed as $y = w^H x$, where x is the received signal vector to be combined, w are the weights computed by the beamforming algorithm and ^{H} denotes transposition and conjugate of $(\cdot)$.

In adaptive antennas design, weights are dynamically calculated with a certain algorithm in order to optimize some signal parameter like signal to interference-plus-noise ratio (SINR), SNR, or BER. An extended variety of algorithms exist in the literature for beamforming purpose and the most appropriated selection is done depending on the signal characteristics of the received signal.

4.3.1 Blind techniques

Blind beamformers make use of an inherent property of the received signal, such as the ciclo-stationarity of the constant modulus. In the latter, the algorithm eliminates the fluctuation of the signal amplitude and computes the weights to minimize the effect produced by those variations. The algorithms that make use of these methods are denoted as Constant Modulus Algorithms (CMA) (Biedka, 2001).

CMA algorithms present an important disadvantage: as the phase information is not considered, the constellation of quadrature phase shift keying (QPSK) signals commonly used in satellite communications appears rotated after beamforming, which imposes the need of an additional phase recovery subsystem in the array output.

4.3.2 Temporal-reference algorithms

Algorithms based on a temporal reference require a known reference included in the frame of the signal, such as training sequences, unique word (UW) or pilot bits. Thus, these schemes are normally used for digital signals. The aim of these beamformers is the minimization of the energy of an error signal integrated by interferences and noise. In order to reduce the order of the problem, the weight calculation is usually done iteratively.

The most popular adaptive filters are the LMS and Recursive Least Squares (RLS) algorithms (Haykin, 2002). Briefly, the main differences lie in the method to calculate and the final convergence behavior: while LMS has a linear complexity order with the number of antennas in the array, RLS makes use of matrix operation, so that the complexity order is quadratic, but the convergence is faster.

An interesting alternative to the LMS is the Normalized LMS (NLMS), which normalizes the adaptive step to avoid variation during the convergence process. The counterpart is the more intensive processing requirements to calculate signal power and normalization operation.

4.3.3 Correlation-based algorithm

In contrast to beamformers based on temporal reference, schemes based on signal correlation do not require the demodulation of any signal. These techniques are the most popular to extract the spatial information for beamforming, and we have focused on the use of the SIMPLE algorithm (Rogstad, 1997). This algorithm has been used by the Deep Space Network (DSN) of National Aeronautics and Space Administration (NASA) to combine the signals received from spatial probes in radio telescopes located in different sites around the Earth surface. The main disadvantage of correlation based schemes is the lack of ability to cancel interference signals.

4.4 Performance comparison

Some simulation comparisons between spatial and blind algorithms are presented to show benefits and drawbacks. Four algorithms have been selected with a 4-element uniform linear array (ULA). The spatial algorithms simulated are post-beamformer interference canceller – orthogonal interference beamformer (PIC-OIB) (Godara, 2004) and minimum power distortionless response (MPDR) (Van Trees, 2002). On the other hand, the blind algorithms are the matrix-free EIGEN and the SUMPLE (Rogstad, 1997). The convergence process is compared as a function of the input SNR as depicted in Fig. 24.

As it can be observed from the above results, spatial algorithms outperform blind ones at low SNR, and vice versa. On the other hand, with medium-low SNR and low or absence of interferences, the behavior of all algorithms is quite similar.

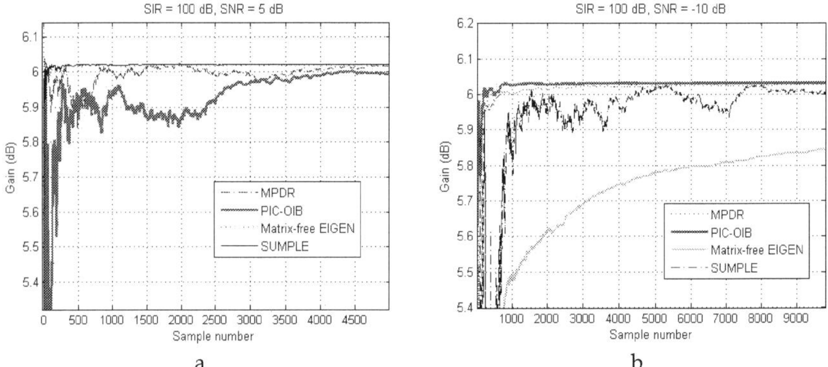

Fig. 24. Convergence behavior of spatial versus blind algorithms in the absence of interferences with several input SNR. a) SNR = 5 dB, and b) SNR = -10 dB.

5. Experimental Test-Bed based on SDR platform

This section presents a test platform known as Adaptive Antenna Array Test-Bed - A3TB, where a comparative study of several beamforming algorithms can be performed and modularity of the architecture is a well proved advantage. The test bed is based on SDR technology and uses a novel architecture that can be used with both blind and spatial-based beamforming algorithms. The A3TB concept can be applied to a number of scenarios as the current version is independent of the signal properties. Simulation results using the A3TB with the APT channel from NOAA satellites show the performance of the concept and the feasibility of the proposed implementation.

The scope of the system development was is to prove the concept of antenna arrays applied to ground stations instead of reflectors for different applications, such as telemetry data downloading or end-user in mobile applications as discussed in the introduction section. In contrast to reflector antennas, antenna arrays offer the possibility of electronic beam-steering avoiding the use of complex mechanical parts and therefore reducing the cost of the antenna. It is also a good chance for Universities and Research Centers aiming to have their own ground station sited in their installations.

5.1 A3TB concept

The A3TB can be defined as *a software-defined radio beamformer applied to a ground station for tracking LEO satellites*. The novelty relies on the use of an antenna array to smartly combine the received signals from the satellite and its implementation based on SDR technology. The reason to use an antenna array instead of a single antenna is to electronically steer the beam in the direction of the satellite along its orbit without requiring a mechanical system for tracking. In addition to the advantages of the use of SDR technology and antenna array, it is the modularity and flexible architecture implemented in the A3TB. Fig. 25 shows the A3TB architecture where it is evident the feasibility to update or change during operation any of the main blocks. It is possible to change during operation the beamforming algorithm and to include new beamforming modules to the system. Furthermore, changes on the BENADC are possible to implement not during operation, but new receiver architecture at off-line such as those options discussed at follow.

In (Salas et al., 2007), the block diagram represents the software system implementation of the first version of the test-bed prototype and most of it is based on VHDL. Depending on the firmware, three options could be installed into the FPGA Virtex4. The option A is implemented with the signal processing on the PC, so the SIMPLE beamforming is done in the module developed in C++. The option B is implemented completely on VHDL and this option need to export the beamforming weights just to draw the array pattern diagram. Finally, in contrast to the option B, the option C is implemented for the LMS beamforming algorithm.

With the first version of the Test-Bed, the modularity on the selection of firmwares was proved switching between A, B or C receiver architectures, and an important result of the Test-Bed development is the hardware resources occupation presented in (Salas et al., 2007).

The advantage of the SDR implementation is that A3TB architecture can be used to process any received signal from a LEO satellite in the appropriate band imposed by the RF stages. Moreover, most of the processing tasks are performed on software, using appropriate routines to process any receive signal. There are 2 main schemes to implement the beamforming stage: SC and FSC [41]. Both schemes are compared in Section 4.2.

The current version of the A3TB in Fig. 25.a was updated to track NOAA satellites in the VHF band, in particular the APT channel. Previous versions of A3TB dealt with LRPT signals from MetOp-A, where a complete receiver with beamforming and synchronization stages has been implemented(Salas et al., 2007; Martínes et al., 2007).

5.2 Implementation of the A3TB

The A3TB prototype consists of 4 main parts as shown in Fig. 25.a. The first part is the antenna array, which has 4 crossed-dipole antennas as depicted in Fig. 25.b. The second part consists of RF-IF circuits which amplify and down convert to IF incoming signals. Furthermore, an automatic gain control (AGC) was implemented using two steps of variable attenuators in the IF domain.

The third part is the SDR platform which consists of the beamforming algorithms implemented on C++ and the FPGA firmware on VHDL, PC and BENADC blocks show in Fig. 25, respectively. The hardware resources occupation for this Test-Bed implementation is similar to one presented in (Martínes et al., 2007). The last part is the software from weather satellite signal to image decoder (WXtoImg) on the PC using the sound card output/input in order to get the weather satellite image.

Since the implemented architecture is FSC the demodulation is not required and the IF signal is digitized. For the signal processing hardware design the BenADC-v4 has been chosen. This solution includes a FPGA Xilinx Virtex4-SX55 with four 12-bit analog inputs at

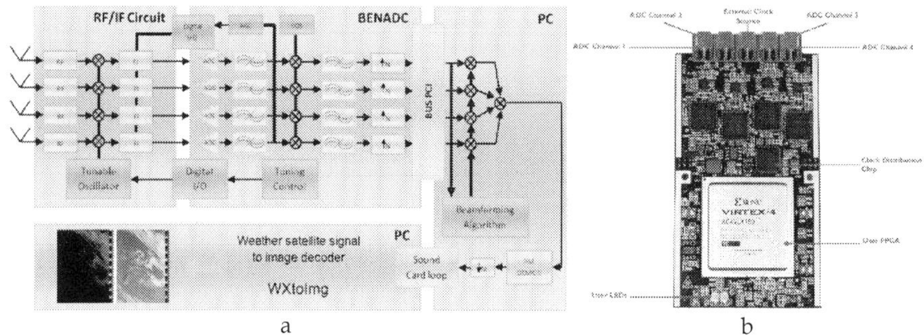

a   b

Fig. 25. a) Block diagram of the A3TB, and b) BenADC – Virtex 4-sx55.

250 Msps (Martínes et al., 2007). Digital samples are transferred to the PC where beamforming and subsequent APT demodulation of the array output are performed using C++ routines. This implementation design offers higher flexibility for testing different beamforming schemes. Finally, demodulated APR frames are sent to the WXtoImg software to show meteorological maps.

The A3TB is controlled by the PC for simulations and field trials. The graphical user interface allows presented in (Salas et al., 2008) the user to choose the beamforming algorithm and set all the parameters of the LEO satellite for tracking such as the number of antennas of the array, distance between the elements, direction of arrival and IF frequency. The C++ routine calculates the beamforming weights and plots the synthesized array factor. Subsequently, the reception of meteorological images has real time system requirements. Thus, it is necessary a data transfer from the FPGA to the C++ module to process the samples continuously, and give APT frames to the audio output of the PC. Since, the meteorological satellites often have a low baud rate, in the case of study with NOAA satellites the data transfer is made using two buffers controlled by a thread.

It is important to mention that the A3TB with SDR architecture can evaluate different beamforming algorithms and receiver schemes. The update of A3TB for larger arrays is immediate, as the basis for algorithms is independent of the number of elements in the array. The architecture of a new ground station concept to track LEO satellites based on software defined radio and antenna arraying as Test-Bed is a well proved choice to evaluate future antenna array architectures for satellite communication and benchmark features of the proposed system. As the A3TB VHF version is based on FSC scheme, the concept can be applied to a number of satellite tracing scenarios.

6. Conclusions

The performance analysis of different beamforming algorithms is an important issue in the new generation antenna array development and research. Thus, A3TB helps to analyze beamforming algorithms paving the way for testing and debugging for posteriori use in larger arrays, such as GEODA. Results obtained in real scenarios with A3TB state, for example, that spatial reference algorithms such as MPDR should be used in the absence of interferences, whereas blind algorithms are appropriate for low SNR conditions. Finally, the A3TB can also serve to validate the performance of calibration procedures.

In future work, the A3TB will deal with the system combining of full modularity with the capability of change firmwares based on the first version design of the Test-Bed, plus the flexible architecture of the current design of the Test-Bed based on VHDL, C++ and Antenna Arraying. Furthermore, the addition of more modules to increase the number of antenna array elements is evident in next generations.

7. Acknowledgment

Authors wish to thank MICINN (Ministerio de Ciencia e Innovación) for grants and CROCANTE project (ref: TEC2008-06736/TEC), INSA (Ingeniería y servicios Aeroespaciales) and Antenas Moyano S.L., for the partial funding of this work. Simulations in this work have been done using CST Studio Suite 2011 under a cooperation agreement between Computer Simulation Technology and Universidad Politécnica de Madrid. Substrates used in prototypes were kindly given by NELCO S.A.

8. References

Torre, A.; Gonzalo, J.; Pulido, M., & Martínez Rodríguez-Osorio, R. (2006). New generation Ground Segment Architecture for LEO satellites. *57th International Astronautical Congress*, pp. 221-226. Valencia, Spain, October 2006.

Tomasic, B.; Turtle, J. & Liu, S. (2002). A Geodesic Sphere Phased Array Antenna for satellite control and communication. *XXVII General Assembly of the International Union of Radio Science,* Maastricht, August 2002.

Godara, L. C. (1997). Application of Antenna Arrays to Mobile Communication, Part II: Beamforming and Direction of Arrival Considerations. *Proc. IEEE ,* vol.85, No.8, (August 1997), pp. 1195-1245.

Martínez, R. & Salas Natera, M. A. (2010). On the use of Ground Antenna Arrays for Satellite Tracking: Architecture, Beamforming, Calibration and Measurements. *61st International Astronautical Congress*, pp. 1-7. Prague, 2010.

Liu, S.; Tomasic, B.; Hwang, S. & Turtle, J. (2006). The Geodesic Dome Phased Array Antenna (GDPAA) for Satellite Operations Support. *IEEE 18th International Conference on Applied Electromagnetics and Communications*, pp. 1-1, Dubrovnik, April 2006.

Tomasic, B. (1998). Analysis and Design Trade-Offs of Candidate Phased Array Architectures for AFSCN Application. *Second AFSCN Phased Array Antenna Workshop,* Hanscom, April 1998.

Sierra Pérez, M.; Torre, A.; Masa Campos, J. L.; Ktorza, D. & Montesinos, I. (2007). GEODA: Adaptive Antenna Array for Metop Satellite Signal Reception. *4th ESA International Workshop on Tracking, Telemetry and Command System for Space Application*, pp. 1-4, Darmstadt, Germany, September 2007.

Arias Campo, M.; Montesinos Ortego, I.; Fernández Jambrina, J. & Sierra Pérez, M. (2010). T/R Module Design for GEODA Antenna. *4th European Conference on Antenna and Propagation,* pp. 115-116, Barcelona, Spain, April 2010.

Franchi, A.; Howell, A. & Sengupta, J. (2000). Broadband mobile via satellite: Inmarsat BGAN. *IEEE Seminar on the Critical Success Factors - Technology, Services and Markets*, pp. 23-30, October 2000.

Evans, J. V. (1998). Satellite systems for personal communications. *Proc. IEEE* , Vol.86, No.7, (July 1998), pp. 1325-1341.

Wu, W. W. (1994). Mobile satellite communications. *Proc. IEEE* , Vol.82, No.9, (September 1994), pp. 1431-1448.

Fujimoto, K. & James, J. (2001). *Mobile Antenna Systems Handbook*. Artech House.

Garcia Aguilar, A.; Inclán Alonso, J. M.; Vigil Herrero, L.; Fernández González, J. M. & Sierra Pérez, M. (2010). Printed antenna for satellite communications. *IEEE International Symposium on Phased Array Systems and Technology*, pp. 529-535, Boston, USA, October 2010.

Geissler, M.; Woetzel, F.; Bottcher, M.; Korthoff, S.; Lauer, A. & Eube, M., (2010). L-band phased array for maritime satcom. *IEEE International Symposium on Phased Array Systems and Technology*, pp. 518-523, Boston, USA, October 2010.

Baggen, R.; Vaccaro, S. & Del Río, D. (2007). Design Considerations for Compact Mobile Ku-Band Satellite Terminals. *EuCAP The Second European Conference on Antennas and Propagation*, pp. 1-5. Edimburg, Scothland, September 2007.

Vaccaro, S.; Tiezzi, F.; Rúa, M. & De Oro, C. (2010). Ku-Band Low-Profile Rx-only and Tx-Rx antennas for Mobile Satellite Communications. *IEEE International Symposium on Phased Array Systems and Technology*, pp. 536-542, Boston, USA, October 2010.

Schippers, H. (2008). Broadband Conformal Phased Array with Optical Beam Forming for Airborne Satellite Communication. *IEEE Aerospace Conference*, pp. 1-17, September 2008.

Kanno, M.; Hashimura, T. T.; Sato, M.; Fukutani, K. & Suzuki, A. (1996). Digital beam forming for conformal active array antenna. *IEEE International Symposium on Phased Array Systems and Technology*, pp. 37-40, October 1996.

Khalifa, I. & Vaughan, R. (2007). Optimal Configuration of Multi-Faceted Phased Arrays for Wide Angle Coverage. *IEEE 65th Vehicular Technology Conference*, pp. 304-308, Boltimore, USA, April 2007.

Salas Natera, M. A.; Martínez Rodríguez-Osorio, R.; Antón Sánchez, A.; García-Rojo, I. & Cuellar, L. (2008). A3TB: Adaptive Antenna Array test-bed for tracking LEO satellites based on software defined radio. *59th International Astronautical Congress*, pp. 313-317, Glasgow. September 2008.

Salas Natera, M. A.; Martínez Rodríguez-Osorio, R. & García-Rojo López, I. (2007). Design of an Adaptive Antenna Array Test-Bed based on Software Radio for Tracking LEO Satellites. *IEEE EuCAP*. Edinburgh, Scotchland, Npvember 2007.

Martínez, R.; Salas Natera, M.; Bravo, A.; García-Rojo, I.; de Haro, L.; Mateo, M. & Gómez, M. (2007). VHF Ground Station with increased angular coverage for reception of meteorological satellites with electronic beamforming. *4th ESA International Workshop on Tracking, Telemetry and Command Systems for Space Application*. Darmstadt, Germany, 2007.

Mailloux, R. (2005). *Phased Array Antenna Handbook,* Artech House, Norwood, Massachusetts, USA.

Josefsson, L. & Persson, P. (2006). *Conformal Array Antenna. Theory and Design*, John Wiley & Sons , Hoboken, New Jersey, USA.

Montesinos, I.; Sierra Pérez, M.; Fernández, J. L.; Martínez, R. & Masa, J. L. (2009). GEODA: Adaptive Antenna of Multiple Planar Arrays for Satellite Communications. *European Conference on Antenna and Propagation*. Berlin, Germany, 2009.

Salas Natera, M. A.; Martínez, R.; De Haro Ariet, L. & Fernández Jambrina, J. (2010). Automated System for Measurement and Characterization of Planar Active Arrays. *IEEE Internationa Symposium on Phase Array Systems and Technology*, pp. 1-6, Boston USA, October 2010.

Sierra-Castañer, M.; Vera-Isasa, M.; Sierra-Pérez, M. & Fernández-Jambrina, J. (2005). Double-Beam Parallel Plate Slot Antenna. *IEEE Transactions on Antennas and Propagation*, vol.53, No.3, (2005) pp. 977-984.

Garg, R.; Bhartia, P.; Bahl, I. & Ittipiboon, A. (2001). *Microstrip Antennas Design Handbook*, Artech House, Norwood, Massachusetts, USA.

Tang, C. & Chen, M. (2007). Synthesizing Microstrip Branch-Line Couplers With Predetermined Compact Size and Bandwidth. *IEEE Transactions on Microwave Theory and Techniques, vol. 55, No.9,* (September 2007), pp. 1926-1934.

Bialkowski, M.; Jellett, S. & Varnes, R. (1996). Electronically Steered Antenna System for the Australian Mobilesat. *IEEE Transaction on Antennas ans Propagation*, vol. 143, No. 4, (August 1996), pp. 347-352.

Alonso, J.; Blas, J.; Garcia, L.; Ramos, J.; Pablos, J. & Grajal, J. (1996). Low Cost Electronically Steered Antenna and Receiver System for Mobile Saltellite Communications. *Trans. IEEE MTT*, vol. 44, No. 12, (December 1996), pp. 2438-2449.

Fernández, J. M.; Rizzo, C. & Sierra-Pérez, M. (2009). Antena Impresa Multihaz con Polarización Circular Derechas/Izquierdas Para Comunicaciones por Satélite en Banda X. *Proccedings of XXIV Simposium Nacional de la Unión Científica Internacional de Radio (URSI)*. Santander, Spain, September 2009.

Encinar, J. A. & Zornoza, J. (2003). Broadband design of three- layer printed reflectarrays. *IEEE Transactions on Antennas and Propagation*, vol. 51, No. 7, (July 2003), pp. 1662-1664.

Padilla, P.; Muñoz-Acevedo, A.; Sierra-Castañer, M. & Sierra-Pérez, M. (2010). Electronically reconfigurable transmitarray at Ku band for microwave applications. *IEEE Transactions on Antennas and Propagation*, vol. 58, No. 8, (August 2010), pp. 2571-2579.

Padilla, P.; Muñoz-Acevedo, A. & Sierra-Castañer, M. (2010). Low Loss 360° Ku Band electronically Reconfigurable Phase Shifter. *International Journal of Electronics and Communications*, vol. 64, No. 11, (November 2010), pp. 1100-1104.

Royce, W. W. (1970). Managing the Development of Large Software Systems. *IEEE WESCON*, pp. 328 – 338, 1970.

Rogstad, D. H. (1997). *The SUMPLE Algorithm for Aligning Arrays of Receiving Radio Antennas: Coherence Achivied with Less Hardware and Lower Combining Loss*. TDA Progress Report, Jet Propulsion Laboratory.

Rogstad, D. H.; Mileant, A. & Pham, T. (2003). *Antenna Arraying Techniques in the Deep Space Network*. Deep Space Communication and Navegation Series, JPL California Institute of Technology, Ref. 03-001, Pasadena, USA.

Biedka, T. E. (2001). *Analysis and Development of Blind Adaptive Beamforming Algorithms*. PdD Thesis, Faculty of Virginia Polytechnique Institute and State University, Virginia.

Haykin, S. (2002). *Adaptive Filter Theory* (4th ed.). Prentice-Hall.

Godara, L. C. (2004). *Smart Antennas* (1st ed.). CRC Press..

Van Trees, H. L. (2002). *Optimum Array Processing. Part IV of Detection, Estimation, and Modulation Theory*. Wiley.